World Without Birds

Written by Nick Lund
Illustrated by Asia Orlando

Workman Kids
New York

FOR ELLIOTT

Text copyright © 2026 by Nick Lund
Illustrations copyright © 2026 by Asia Orlando

Hachette Book Group supports the right to free expression and the value of copyright. The purpose of copyright is to encourage writers and artists to produce the creative works that enrich our culture.

The scanning, uploading, and distribution of this book without permission is a theft of the author's intellectual property. If you would like permission to use material from the book (other than for review purposes), please contact permissions@hbgusa.com. Thank you for your support of the author's rights.

Workman Kids
Workman Publishing
Hachette Book Group, Inc.
1290 Avenue of the Americas
New York, NY 10104
workman.com

Workman Kids is an imprint of Workman Publishing, a division of Hachette Book Group, Inc.
The Workman name and logo are registered trademarks of Hachette Book Group, Inc.

Design by Daniella Graner

The publisher is not responsible for websites (or their content) that are not owned by the publisher.

Workman books may be purchased in bulk for business, educational, or promotional use.
For information, please contact your local bookseller or the Hachette Book Group
Special Markets Department at special.markets@hbgusa.com.

Library of Congress Control Number: 2024053395
ISBN 978-1-5235-1802-9
First Edition January 2026

Printed in Dongguan, China (10/25) TLF on responsibly sourced paper

Cover © 2026 Hachette Book Group, Inc.

10 9 8 7 6 5 4 3 2 1

CONTENTS

Introduction / Hopefulness...v

1. Our World of Birds, and a World Without Them...1

2. A Look at the Ways Humans Are Causing Extinction...7

3. Temperate Forests...24
   In a World Without Birds: Seed Dispersal and Pollination...35
   Overhunting and the Extinction of the Passenger Pigeon...41
   Pesticides and the Amazing Recovery of the Bald Eagle...43

4. Tropical Forests...46
   In a World Without Birds: Insect Apocalypse...54
   Climate Change and the Extinction of the Po'ouli...58
   The Pet Trade and the Survival of the Hyacinth Macaw...61

5. The Poles...64
   In a World Without Birds: Birds as Prey...73
   Overexploitation and the Extinction of the Great Auk...77
   Rodents and the Incredible Recovery of the South Georgia Pipit...80

6. Deserts..84
In a World Without Birds: Carrion Cleaners.....................92
Human Encroachment and the Extinction
of the Arabian Ostrich...96
Introduced Cats and the Survival of the Night Parrot.......99

7. Islands..102
In a World Without Birds: Empty Ecosystems..................111
The Discovery of Extinction and the Dodo....................115
Island Development and the Survival of the Kagu............118

8. The Open Ocean......................................122
In a World Without Birds: Economic Impact..................131
Introduced Livestock and the Extinction
of the Guadalupe Storm-Petrel...............................135
The Feather Trade and the Unlikely Survival
of the Short-Tailed Albatross................................138

9. Fresh Water..142
In a World Without Birds: Cultural Loss.....................150
Pollution and the Extinction of the Colombian Grebe..........154
Agricultural Conversion and the Survival
of the Crested Ibis..156

10. What Can We Do?..............160

INTRODUCTION
HOPEFULNESS

Before we get into anything else, I want you to know that this is a book about hope. It's about a lot more than just that—it's also about stupidity and greed and ignorance and love and sadness and joy—but it's mostly about hope. And birds.

Birds are a symbol of hope. Flight gives many birds their freedom and allows them to soar above difficult circumstances on the ground. If they don't like how things are going in one place, they can pick up and fly somewhere better to start over. Birds have used their wings to find their way into every corner of this planet, from the ice-capped poles to the densest jungles to the baking-hot deserts.

The problem is, humans have made something of a mess on Earth, so some birds have few places left to go. There's never been a species like us: We've evolved to the point where the normal rules of evolution don't apply to us anymore.

WORLD WITHOUT BIRDS

Our large brains have allowed us to grow all the food we need, build massive cities to live in, and travel anywhere on Earth in a day. We've transformed the planet to help our species grow, but given little thought along the way to all the other species with which we share the planet.

We've cut down forests, filled in wetlands, carved up grasslands, polluted the oceans, laid millions of miles of roads, and built millions of buildings. We've hunted and fished for thousands of species of animals and cleared billions of acres to plant crops. Each of these habitats—forests, wetlands, grasslands, oceans, and many more—were filled with plants and animals that had evolved over millions of years to live there. They were just in our way.

No creature has been spared, not even birds. Scientists say that North America has lost one in every four individual birds, more than 3 billion total individuals, in just the past fifty years. According to the International Union for the Conservation of Nature, humans have caused the extinction of 159 species of birds since 1500, with more than forty additional species either extinct in the wild or presumed extinct. After thriving for more than 60 million years on Earth, birds are being wiped out by humans in just a few hundred.

INTRODUCTION

And yet, there's hope.
I promise there is.

There's hope in the fact that unlike asteroids and volcanic eruptions, we humans can stop ourselves. Our large brains got us into this mess by giving us the ability to cause profound change to the planet, but our intelligence has also given us the ability to recognize and understand our impacts—and heal them.

Those are the two steps humans must take to prevent extinction: recognize our impacts and change our behavior.

Thankfully, we've got most of the first step covered. We know more about the world than any other species ever could, and that includes a detailed and evolving understanding of the impacts we're causing. We know a lot about the different kinds of life on Earth and what conditions they need to survive, and the various ways our activities are harming life. We know, with more clarity than ever before, exactly how we're putting species at risk.

It's the second part, changing our behavior, that's the tricky one. Changing behavior is difficult—just ask anyone who has tried to stop biting their nails or eating too much junk food. (Actually, you can just ask me about those things: It's hard.) Now imagine we've got to get *all the humans on Earth* to change their behavior. It's daunting, and there's always resistance from those who don't want to change or whose livelihoods depend on a particular way of doing things. I know that I need to eat better but I don't always do it, and humans know that we need to live better and make less of an impact on the Earth but we don't just change.

Except when we do. Though change is hard, there are lots of examples of humans changing their behavior in order to protect the Earth and protect individual species. We can do it when we want to, we have done it before, and we're continuing to do it now. This book exists to celebrate birds in all their diversity, and then try to understand the potential impacts of their disappearance. We'll explore the various ways in which birds are under threat, but also the courageous work being done to save them. These celebrations and questions and history lessons are important to remind us about what we're fighting for, and what is possible if we fail. Together we can protect this planet if we have hope.

WORLD WITHOUT BIRDS

I bet you've seen a bird today. **Somewhere, at some point, I bet you have. All it really takes is a glance out the window. In any city or country most anywhere in the world, look outside and you could see a bird. I'd also bet that you *haven't* seen a tiger, or a bear, or a mouse, or an eel, or a trout, or a praying mantis, or a salamander, or a snake.**

Birds are *everywhere*. Birds are the most visible and evident wild creatures on Earth because they live all over the Earth: from cities and towns to the remotest forests, deserts, mountains, and oceans. In the 65 million years since the mass extinction that killed the dinosaurs, birds have conquered the globe. Over those millennia, birds have evolved into many thousands of different species, with more than 10,000 species alive today.

Flight has been perhaps their most important adaptation. It has allowed them to quickly cover the globe, and no land mass—from the centers of the largest continents to the loneliest rocky islets on the open sea—is free of birds. Flight allows birds to escape predators and range easily over huge areas, unlike other large animals. The protection of flight has given birds confidence that other animals don't have: the confidence to have bright colors and sing loud songs,

Blue Jay

WORLD WITHOUT BIRDS

I bet you've seen a bird today. **Somewhere, at some point, I bet you have. All it really takes is a glance out the window. In any city or country most anywhere in the world, look outside and you could see a bird. I'd also bet that you *haven't* seen a tiger, or a bear, or a mouse, or an eel, or a trout, or a praying mantis, or a salamander, or a snake.**

Birds are *everywhere*. Birds are the most visible and evident wild creatures on Earth because they live all over the Earth: from cities and towns to the remotest forests, deserts, mountains, and oceans. In the 65 million years since the mass extinction that killed the dinosaurs, birds have conquered the globe. Over those millennia, birds have evolved into many thousands of different species, with more than 10,000 species alive today.

Flight has been perhaps their most important adaptation. It has allowed them to quickly cover the globe, and no land mass—from the centers of the largest continents to the loneliest rocky islets on the open sea—is free of birds. Flight allows birds to escape predators and range easily over huge areas, unlike other large animals. The protection of flight has given birds confidence that other animals don't have: the confidence to have bright colors and sing loud songs,

Blue Jay

OUR WORLD OF BIRDS, AND A WORLD WITHOUT THEM

and the confidence to live among humans. Many earthbound creatures depend on camouflage to avoid predators, or only come out under cover of darkness, but not birds. The safety of flight gives them the freedom to show themselves.

The power of avian flight astounds scientists and birders alike. The four-ounce arctic tern has been recorded flying more than 44,000 miles between Greenland and Antarctica each year, the longest migration of any animal on Earth. Ruppell's griffon vultures have been recorded at altitudes of more than 36,000 feet, the same height as commercial airliners. Peregrine falcons have been recorded traveling at 242 miles per hour during their hunting dives, the fastest creature on the planet.

Peregrine Falcon

And some birds don't fly at all! They have evolved to live all kinds of lifestyles, including those where they've lost the need to fly. Ostriches, the biggest birds in the world, are completely flightless but can run up to sixty miles per hour to escape predators. Penguins evolved to hunt fish, and lost the ability to fly as their bodies were streamlining to become agile underwater torpedoes. There were no mammals on the islands of New Zealand that broke away from what is now Australia 80 million years ago, so when birds flew across the

Tasman Sea, they found a land without mammalian predators. Some evolved to fill roles occupied elsewhere by mammals, and many of these birds, like the kiwi, the kākāpō, and the giant moa, lost the ability to fly.

Evolution has driven birds into all kinds of shapes and sizes, and each living species is unique. Raptors like hawks, eagles, owls, and falcons are fearsome predators, using stealth, speed, and power to bring down prey. Shearwaters, albatross, and petrels are long-winged seabirds that spend their entire lives sailing over the open oceans, coming to land just once a year to breed. Pelicans trap fish in their giant netlike beaks. Kingfishers plunge face-first into water to snatch fish. Crossbills evolved their namesake overlapping mandibles to help them pry open pine cones in search of seeds inside. There's too much diversity to cover; I'd need a thousand books!

The diversity and everywhere-ness of birds have endeared them to humans like perhaps no other creatures. Birds were likely most important to early humans in one of the same ways they are most important to us today: as food. Birds and their eggs were a source of fresh meat, and chickens, geese, and other birds were some of the earliest domesticated animals. Chickens remain one of the most important sources of food on the planet, with at least 25 billion of them alive at any given time, producing more than *1 trillion eggs per year.*

But birds have inspired humans in many other ways. Images of birds appear in artwork created thousands of years ago in Australia

OUR WORLD OF BIRDS, AND A WORLD WITHOUT THEM

and Europe. Bird-shaped deities were worshipped by dozens of different civilizations, including the ancient Egyptians, ancient Greeks, and Indigenous American tribes. Bird feathers decorated ceremonial clothes across the globe, and birds have long been kept as pets to delight their owners with their beautiful songs. Birds decorate the national flags of at least twenty countries, more than any other animal. We name our sports teams after them.

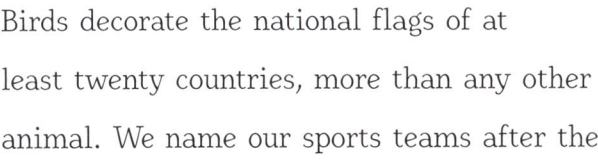

Domestic Canary

We say people are as *graceful as a swan*, or have *eagle eyes*, or are just as *free as a bird*. When we're up early it's because *the early bird catches the worm*, but when we're up late we're *night owls*. Birds have influenced every aspect of our speech, culture, and art.

And yet, for most of us, the joys of birds are more commonplace. It's the flash of color as a northern cardinal visits our backyard feeder. It's the almost incomprehensible display of coordination of a murmuration of European starlings at dusk, or the impressively ominous megaflock of American crows headed to roost. The cheery song of an American robin on a spring morning, or the melancholic wail of a common loon. Birds can be obvious reminders that we live in a wild world, and are a welcome intrusion into our civilized lives.

A world without birds, then, would be a world without these joys. It would be a quiet place, and a lonely place, and humanity would lose one of its last strong connections to the natural world.

CHAPTER TWO
A LOOK AT THE WAYS HUMANS ARE CAUSING EXTINCTION

We have been warned about birds going extinct and humans causing a sixth mass extinction. Scientists are learning more and more each day, and with increasing exactitude, about the scale and scope of our environmental harm. And they're telling us what we need to know to save birds, if we only listen.

A clear assessment came from a group called the Intergovernmental Science-Policy Platform on Biodiversity and Ecosystem Services, helpfully shortened to IPBES. The group, associated with the United Nations, has official involvement from more than 130 countries, along with thousands of nonprofit environmental organizations, research institutions, and scientists.

The IPBES tries to understand how the planet is doing, and then tells us about it. They met those goals in their latest report, from 2019, and presented a clear and concise statement of this critical moment for birds and the environment. The key finding was highlighted in bold:

"1 million animal and plant species are now threatened with extinction, many within decades, more than ever before in human history."

THE IPBES OUTLINED FIVE MAJOR DRIVERS OF CHANGE IN NATURE:

- **CHANGES IN LAND AND SEA USE—HOW HUMANS ARE CHANGING THE PHYSICAL LANDSCAPE:** CLEARING FOREST, GRASSLANDS, WETLANDS, AND OTHER NATURAL AREAS TO MAKE WAY FOR HOMES, AGRICULTURAL FIELDS, AND OTHER HUMAN USES.

- **DIRECT EXPLOITATION OF ORGANISMS:** THE KILLING OF ANIMALS, PLANTS, AND OTHER ORGANISMS, MAINLY THROUGH HARVESTING, LOGGING, HUNTING, AND FISHING.

- **CLIMATE CHANGE:** THE LONG-TERM SHIFTS IN TEMPERATURE AND WEATHER PATTERNS CAUSED BY HUMANS ADDING CERTAIN CHEMICALS TO THE ATMOSPHERE.

- **POLLUTION:** CHEMICALS HUMANS ARE ADDING TO THE AIR AND WATER.

- **INVASIVE ALIEN SPECIES:** ORGANISMS THAT HUMANS INTRODUCE INTO NEW ENVIRONMENTS WHERE THEY CAN CAUSE DISRUPTION AND HARM.

There you have it. These five factors are responsible for all or nearly all the human-caused bird extinctions on Earth. If they're left unchanged, these five factors could result in a world without birds.

It's scary that so much destruction can be captured in such a short list, but it's important for humans to be aware of what we're doing. We can only begin to fix the problems we're causing if we know what the problems are.

We didn't always know we had the power to destroy the world.

For most of human existence it was inconceivable that we could cause other species to go extinct. The concept of extinction at all—the idea that animals could just disappear from the planet and be gone forever—was revolutionary. Some religions, for example, believed that the human world was separate and distinct from the natural world, and that all had been created by a god or gods. Humans simply couldn't cause another of God's creatures to disappear. At the same time, some religions taught that the world was made for humans to conquer and subdue, and that the value in natural resources was in their usefulness to us and their exploitation.

Early humans were causing extinctions well before we had a word for it. For at least 1.5 million years, the advanced hunting techniques of our ancestors were enough to overwhelm the populations of large

A LOOK AT THE WAYS HUMANS ARE CAUSING EXTINCTION

mammals and contribute to their extinction, though we couldn't comprehend the consequences of our actions until much later. Humans arrived in North America sometime around 12,500 years ago, and within a few thousand years nearly 70 percent of the region's megafauna was extinct, including mammoths, mastodons, camels, and giant sloths—along with many of the carnivores that preyed on those species, including lions, American cheetahs, short-faced bears, and saber-toothed cats. We made it to South America about 10,000 years ago and quickly dispatched armadillos the size of cars, more giant sloths, and huge rodents. The same pattern followed human expansions into Australia, Europe, Asia, and anywhere else we arrived. Scientists continue to debate the role humans played in these extinctions relative to other factors at the time—namely a changing climate as glaciers receded—but overhunting from humans was certainly a contributor.

Our thinking about extinction, and the human relationship to the natural world, changed radically in the 1800s. The most important change was the dawning understanding of evolution. Many naturalists, philosophers, and thinkers were coming to terms with new scientific discoveries during the Industrial Revolution of the 1800s, and some, like Alfred Wallace, Joseph Hooker, Thomas Huxley, and most famously Charles Darwin were working on a new understanding of how different species on Earth came to be.

The accepted idea in Europe at the time was that the various species of plants and animals on Earth were placed there, perfectly formed, by God. Darwin's theory of evolution was radically different: Species of life on Earth were constantly changing in order to survive.

Unlike the popular understanding of the time that the natural world was a paradise, a Garden of Eden, Darwin acknowledged that it was actually a hard, dangerous, and not particularly kind place in which to live. Wild species are in a constant struggle for survival: to eat, to not be eaten, to reproduce. Individuals of the same species struggle against each other for territory and for mates, and different species struggle against one another for dominance or survival.

The theory of evolution recognized that breakthroughs in these struggles occur when there are slight variations between parents and their offspring. Darwin understood that when living creatures reproduce—create seeds, have babies, etc.—their offspring aren't exactly the same as their parents. Some offspring might be a little faster or slower, or have longer necks, or any number of other differences. Most of these variations have no impact at all on the animal, but some might help it survive. A slightly faster animal might live through an

A LOOK AT THE WAYS HUMANS ARE CAUSING EXTINCTION

attack from a predator where its slower relative might not, or a longer-necked animal might be able to reach food others can't. These animals might then pass their variations onto their own offspring, and over many, many generations, the entire species might change to become faster, or longer necked, or whichever modification helped that animal survive in a harsh and constantly changing world.

Common Cactus-Finch

Darwin's ideas were considered radical not only because they contradicted the popular religious understanding of the world at that time, but because they recognized the interconnectedness of life. The world was shocked at the idea that humans weren't "special" beings, separate and better than the rest of the natural world, but rather the result of the same evolutionary processes that developed every other living thing on Earth.

At the same time, life was fragile. While some species won the battle for survival, others lost.

Implicit in Darwin's theories was the fact that the history of life on Earth is riddled with species that didn't survive. Extinction was real.

It was an idea that helped find an answer to what was then one of the world's biggest mysteries: fossils. Humans had for thousands of years known about fossilized bones that were similar in some ways to certain modern animals but weren't from modern animals. No one knew exactly what they were. Did they belong to dragons? Giants? Were they planted there by God or the devil? Evolution helped provide an answer. The bones belonged to creatures that had gone extinct.

Our understanding of evolution and extinction was growing at the same time we were also beginning to recognize the impact we were having on the world. Thousands of factories sprung up during the Industrial Revolution, many of which began spewing pollution into the air and water. Dense cities grew around industrial centers, adding to the mess.

Many in the 1800s believed in the concept of "nature's bounty," that nature was so powerful that it was indestructible. However, some began to notice that populations of wild animals were decreasing. Of all the creatures on Earth, fish, out there in the boundless sea, were believed to be endlessly plentiful. Fishing technology improved in the 1800s, meaning that fishermen who once had to use sail power to get around could move much faster and more powerfully with steam engines, allowing them to drag larger nets

> Charles Dickens painted a dire portrait of pollution in London in his 1853 novel *Bleak House*: "Smoke lowering down from chimney-pots, making a soft black drizzle, with flakes of soot in it as big as full-grown snowflakes—gone into mourning, one might imagine, for the death of the sun."

A LOOK AT THE WAYS HUMANS ARE CAUSING EXTINCTION

deeper than ever before. Many fishing communities began to notice that they had to move farther offshore to find economically vital fish species like Atlantic cod, Atlantic herring, and others, and that the fish were smaller than in the past. We were realizing that there were limits to even the most plentiful of resources.

As the human footprint expanded, we began to notice the destruction of wild areas. The phenomenon was perhaps most evident in America, a relatively new and wild land. Unlike Europe, which had been densely occupied by humans for thousands of years, America in the eighteenth century was still sparsely occupied, especially west of the Appalachian Mountains. That changed rapidly, as gold, free land, timber, oil, and other natural resources drew hundreds of thousands of people west. Indigenous people were ruthlessly killed or forced off their lands. The American bison, a plentiful icon of the West, was hunted without conscience, their numbers reducing from an estimated 40 million in 1830 to just 300 in 1900. Wild rivers full of salmon were being dammed, millions of acres of prairie converted into fields, and millions more acres into pasture.

Wild America was being erased, but some people were beginning to put up a fight.

WORLD WITHOUT BIRDS

A movement was growing in the mid-1800s that aimed to conserve nature instead of using it all up. The relative lack of private land ownership and development in the western United States meant that huge areas of land could potentially be protected by federal law. (Of course, Indigenous peoples had been living in these lands for centuries and were removed by force, an ongoing, evil chapter in American history.) The federal government established the Department of the Interior in 1849 to manage and conserve the nation's natural resources and federal lands.

A few decades later, the nation hit on what's been called its "best idea." For decades, reports of an especially beautiful landscape, filled with gushing geysers and multicolored rock formations, had reached the public's ears from expeditions that traveled through what is now northwestern Wyoming. Subsequent visitors marveled at the unique geology of the area and began to advocate for the land to be set aside, protected from loggers and agricultural development. In 1872 President Ulysses S. Grant signed the law creating Yellowstone as America's first national park. Dozens more parks would follow over the next century, as well as many other types of protected lands, including state parks, national monuments, and national wilderness areas (the highest level of protection, with no mechanized travel permitted).

> Today the National Park System is made up of 433 individual units encompassing more than 85 million acres in all fifty states and the District of Columbia, covering about 3.4 percent of the nation.

A LOOK AT THE WAYS HUMANS ARE CAUSING EXTINCTION

The purpose of the National Park Service, formally established in 1916, was to "conserve the scenery and the natural and historic objects and the wild life therein and to provide for the enjoyment of the same in such manner and by such means as will leave them unimpaired for the enjoyment of future generations." That sentiment represented a mission statement not just for the park service but for the entire conservation movement. We should not deprive future generations of the natural world we enjoy.

But the shift in perspective didn't solve our march toward extinction. In fact, humanity continued to discover more ways it was killing off the world around it. Pollution was our next awakening. For decades, humans had built factories around the world without concern or understanding of how to properly dispose of the often toxic waste they produced. The answer was simply to just get rid of it. Smoke from coal-fired power plants, factories, and gasoline-powered vehicles added massive amounts of untreated chemicals to the air. Nitrogen oxides, sulfur dioxide, carbon monoxide, suspended particulate

Sometimes it took large-scale environmental disasters to wake the nation up to the impacts of pollution. One was the Donora Smog of 1948, which saw a specific weather pattern trap toxic smoke from a zinc factory close to the ground for five days, causing respiratory problems for thousands of residents of Donora, Pennsylvania, and killing twenty. "It was not until that tragic impact of the Donora that the Nation as a whole became aware that there might be serious danger to health from air contaminants." —Leonard Scheele, surgeon general, 1949.

17

matter, and many other chemicals were released into the air without regulation.

Waterways were similarly degraded. There are many sources of water pollution, including the direct discharge of municipal sewage and industrial waste into the water, or indirect pollution caused by rainwater washing over or through contaminated surfaces on its way to collect in a larger body. The environmental and recreational impacts are often immediate and obvious, like discolored, fetid, or putrid water unfit for fishing or swimming. Water polluted by sewage or other contaminants can cause diseases like cholera and typhoid. Passage of the Clean Water Act in 1972 and the Safe Water Drinking Act of 1974 helped reduce pollution levels in American waterways, though water pollution remains a major problem in the United States and around the world.

Invasive species may not be the most obvious threat, but they've had a massive impact on our environment. Life evolves in response to the conditions around it, and introducing a new species into an ecosystem and changing its conditions can have major consequences. For example, birds living in places with lots of predators need to be wary to survive, and have evolved defenses like camouflage accordingly. However, birds living in places without

> Birds and other wildlife were at the center of another national water-pollution tragedy: the 1989 Exxon Valdez spill. The oil tanker ran aground in southern Alaska and spilled more than 11 million gallons of oil into the water, killing seals, fish, otters, whales, and approximately 250,000 seabirds. For many alive at the time, images of oil-drenched birds dead or dying on beaches are unforgettable.

A LOOK AT THE WAYS HUMANS ARE CAUSING EXTINCTION

natural predators, like on remote islands, don't need those defenses, and often evolve to become incautious or unafraid.

So when humans introduce a predator to an island, such as a fox or rat or cat, the native bird species are wholly unprepared to defend themselves.

Invasive species cause harm in less obvious ways, too, like introduced plants outgrowing native species and causing food or habitat loss for wildlife, or nonnative birds taking over the nesting holes of native species. Once thriving in a new ecosystem, nonnative species can become widespread and entrenched, making them difficult or impossible to remove. Humans have either accidentally or intentionally introduced new species to ecosystems all over the globe, often with disastrous consequences, and are just now understanding the full breadth of their impacts.

Climate change is the latest human-caused environmental crisis we've begun to acknowledge. Climate is the average weather that exists in a place for many years. The climate in Antarctica is frigid and dry; the climate in the Amazon rainforest is hot and wet; the climate in New York City is hot in the summer and cold in the winter.

The term "climate change" describes long-term shifts in climate. As in—what if New York City wasn't cold in the winter? Or what if the snow in Antarctica melted? It may seem far-fetched, but the Earth's climate is constantly changing—Antarctica was covered in rainforest 90 million years ago. However, this change typically happens over such a long period of time that it's not noticeable over the course of a single lifetime. Climate change can have a drastic impact on life. Species evolve to inhabit specific ecological niches, and if the climate changes such that those niches change, the species either have to rapidly adapt, or perish.

Now humans are causing the world's climate to change much faster than normal, and faster than many species can adapt.

While climate change is the latest of our environmental impacts we've recognized, it might be the most dangerous of all.

Rapid or drastic climate changes can be deadly. Naturally occurring climate changes have played a leading role in many extinctions, from the five past mass extinctions to thousands of little-known events throughout the long history of life on Earth. Plants were quickly colonizing land 375 million years ago, and their rapid intake of carbon dioxide changed the Earth's climate, cooling it down and contributing

A LOOK AT THE WAYS HUMANS ARE CAUSING EXTINCTION

to a mass extinction that killed nearly 80 percent of life. A series of large volcanic events toward the end of the Triassic period spewed gases into the atmosphere, warming the planet, changing sea levels, altering the chemical composition of the oceans, and ultimately killing more than half the species on Earth. Climate change has always been deadly.

The climate is changing once again, but this time, the effects are being intensified and quickened by humans. Ever since the Industrial Revolution, humans have been responsible for the release of massive amounts of greenhouse gases that have the effect of trapping heat in our atmosphere. More and more heat is trapped as we continue to emit greenhouse gases, and eventually our climate warms.

There are natural sources of greenhouse gases, like wildfires and volcanoes, but humans are releasing much more than what occurs naturally. Humans produce greenhouse gases by burning coal, oil, and natural gas for power; by burning gasoline to drive cars; and by raising more than 1.5 billion cattle. We also keep more greenhouse

Believe it or not, cows and other livestock release methane, a powerful greenhouse gas, in their burps and farts. Scientists estimate that about 14 percent of all human-induced climate change comes from farm animal "emissions."

gases in the atmosphere by cutting down millions of acres of forest that would otherwise help pull carbon dioxide out of the air through photosynthesis.

Human release of greenhouse gases isn't as noticeable as, say, gigantic volcanic eruptions, or even pollution dumped into a river. It took human scientists until the 1970s and '80s to recognize and understand the threat, and several decades more for widespread acceptance and understanding of the problem. In the meantime, humans are causing the planet to warm at a rate ten times faster than at any time in the past 65 million years, and the effects are being felt already.

As our climate changes and causes ecosystems to change, birds will need to move, adapt, or perish. Rising ocean temperatures are changing the places where certain fish live, forcing seabirds to leave their traditional nesting islands to find them. A warming planet causes ice caps to melt and then sea levels to rise, inundating coastal marshes, beaches, nesting islands, and other important areas for birds. The migrations of many bird species are triggered by temperature, and a changing climate means that birds are leaving at different times. Temperature and precipitation

American scientists have identified nearly 400 bird species, about two-thirds of all American bird species, that are at risk of extinction from rising temperatures.

A LOOK AT THE WAYS HUMANS ARE CAUSING EXTINCTION

help define the borders of different ecosystems, but habitats are changing with changing temperatures. Forests once kept in check by frigid temperatures are creeping northward, encroaching on tundra nesting habitats utilized by billions of waterfowl, wading birds, and others. The impacts are global, and they're happening now.

But, again, humans are waking up to our impacts. We know what we're doing that is causing the climate to change: releasing greenhouse gases. Reducing our emissions would slow the pace of change, and we can do that by getting electricity from renewable sources like solar and wind instead of burning coal or oil. We can also reduce emissions by driving electric cars instead of those powered by gasoline. We can be more efficient in our energy use. We can plant forests to take carbon dioxide out of the air. We know about these solutions; now we just need to implement them.

You did it.

If you're still here, that means you've gotten through the hard part. Humans have done a lot of pretty destructive things to birds and other wildlife. We're still doing many of those things. We're doing them all over the world—in forests, in deserts, at the poles, and on the open ocean—but we know what they are, and we can stop them. If we want to prevent birds from going extinct, we need to look our mistakes in the face and fix them. In some cases, it's too late; in many cases, it's not.

PILEATED WOODPECKER

BLACK-AND-WHITE WARBLER

CAROLINA CHICKADEE

Sandwiched between tropical forests near the equator and cold, pine-dominated boreal forests closer to the poles, **temperate forests ring the globe in both the northern and southern hemispheres. Temperate forests are something of a happy medium. They feature the lengthy, hot summers of tropical forests but also the changing seasons of the boreal forests. They're home to thousands of bird species, some of which are seasonal visitors and others year-round residents.**

Temperate forests are full of trees, of course, and those trees provide everything for birds. Hundreds of tree species—oak, ash, maple, fir, beech, southern beech, birch, poplar, elm, pine—and thousands more plant species in the undergrowth all supply something that birds need to thrive.

Warblers, vireos, tanagers, and many other songbirds feast themselves on caterpillars emerging to eat leaves in the spring and summer. Woodpeckers use their bills to hammer into tree bark looking for insect larvae hiding inside. Flycatchers and kingbirds wait patiently on branches for butterflies, moths, and other insects to flutter by.

In winter, when bugs are scarce, trees provide seeds and fruit for birds to eat. Finches and grosbeaks can cover thousands of miles in search of trees with the best crop, then crush the tough seeds with their oversize bills. Chickadees, nuthatches, and tits hide hundreds

TEMPERATE FORESTS

of seeds in tree bark and other crevices, ensuring they have a reliable supply all winter long. The trees provide the birds with homes, too. Branches give safe places for birds to build their nests, and natural holes in the trunks are perfect nooks for owls, nuthatches, chickadees, titmice, bluebirds, woodpeckers, and even some parrots.

Pileated Woodpecker

Temperate forests cover about a quarter of all land on Earth, providing habitat to thousands of species along with incalculable environmental benefits. But many of the same attributes that make temperate forests a welcome home for birds also make them attractive to humans. The planet's temperate zones are some of the most populated, meaning they're some of the most impacted environments on the planet. For thousands of years, humans have cleared temperate forests for timber, for agriculture, and to build towns and cities. They've hunted in the forests for food, polluted the air and water, and generally affected these lands as much as anywhere on Earth.

The various human impacts have caused overall bird populations to decline steadily. The National Audubon Society estimates that North America has lost 3 billion birds since 1970. What would these forests look like—and sound like—without birds?

The most remarkable feature of most temperate forests is their changing seasons. As the tilted Earth revolves, the temperate forests are exposed to different amounts of sunlight throughout the year. The forests are warm and the days are long when they are tilted toward the sun, and the forests are colder with shorter days when tilted away: summer and winter, with spring and fall in between.

The seasonality of temperate forests dictates the lives of the birds found within. To handle the major swings in temperature between summer and winter, many bird species *migrate*—fly hundreds or thousands of miles to areas with better weather. The annual north-south migration of birds to and from temperate forests is one of the most incredible animal spectacles on Earth.

Spring: Bloom and Boom

The temperate forests really come alive in spring and summer, and billions of migratory birds arrive to join the party. Rising temperatures and increasing daylight in early spring trigger trees and plants to begin growing, and the spring forest bursts with greenery. Insects of all kinds appear to eat the new leaves and begin their life cycles. Birds love to eat insects, and this sudden bounty is

just what birds need to help their babies grow big and strong. Most birds in these forests lay their eggs in spring to take advantage of the summertime insect buffet.

Thousands of bird species spending their winters in the tropical forests near the equator begin to get antsy in spring. Their instincts are telling them it's time to move north. They eat nonstop to store extra fat for the arduous journey ahead. Then, on some calm night, they take off and head north.

The majority of these birds migrate at night, following the stars and relying on instinct to guide them to the areas where they were born. It is a monumental journey. More than two dozen species of warbler in North America, each bird weighing less than an ounce, fly about 6,000 miles in the spring to get to the insect-rich northern forests. The ruby-throated hummingbird, one of the smallest birds in North America, flies from the forests of Central America up through Mexico and nonstop across the Gulf of Mexico to Texas or Louisiana, where it then flies thousands of additional miles north to breed.

It's a difficult journey under ideal conditions, but humans are making migration—and life in the temperate forests—harder and harder. Temperate forests cover large areas of land, and their relatively tolerable climates—warmer than the far north or south and cooler than the tropics—have made them attractive places for humans to live. Hundreds of thousands of square miles of temperate forest have been cut to make way for cities, towns, farm fields, and other human development. Fewer forests mean fewer places for migratory birds to lay their eggs or stop to refuel during their migrations.

Plus, all that human development is disorienting and dangerous. Birds migrated for millions of years using stars in the clear sky to guide them, but humans have completely changed how the world looks at night in just over one hundred years. Now the landscape is aglow with human lighting, obscuring the stars and disorienting birds. The buildings that pepper the ground are covered with windows made of glass that trick birds by reflecting the sky or vegetation. Scientists estimate that about 1 billion birds die in North America each year after colliding with glass windows, and still more die from colliding with vehicles, transmission lines, and other structures.

But the trip is still worth it for birds who want to raise their young on the bounty of insects that emerge in spring. Insects, especially caterpillars, are full of protein, and nearly all birds in the temperate forest rely on insects for food. Even birds that mainly eat different foods during other times of the year, like the seed-eating finches or nectar-drinking hummingbirds, feed insects to their young. Over millennia, birds in the temperate

TEMPERATE FORESTS

forests have timed their breeding seasons to coincide with the spring insect emergence.

It's a good thing, too, because baby birds are hungry. Studies have shown that Carolina chickadees, a fist-size bird common throughout the American southeast, need between *6,000 and 9,000* caterpillars to raise a single clutch of their young. It's the same story across the globe, where the thousands of species of warblers, vireos, sparrows, buntings, cuckoos, thrushes, pipits, grosbeaks, orioles, tanagers, blackbirds, swifts, swallows, flycatchers, and many more fly thousands of miles to raise young in temperate forests.

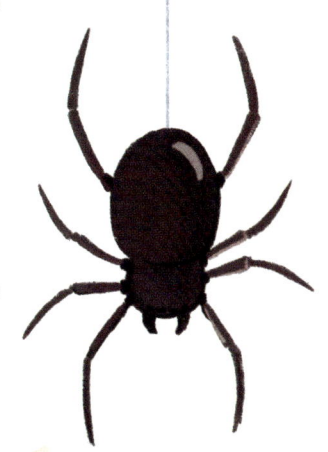

Though we think of hummingbirds as primarily feeding on nectar from flowers, their diet is up to 80 percent insects and spiders.

And that's what the spring and summer are about for birds in the temperate forest: laying eggs and raising young while there's food to be had. They need to work quickly, because the good times don't last forever. After just a couple of months, birds begin to feel the seasons change once again. The days are getting shorter and colder, and the birds instinctively know that there won't be insects around for much longer. The newly hatched young prepare themselves, and on calm nights throughout the late summer and autumn, billions of birds take to the skies to migrate once again, this time south to the warm tropical forests where they began their year.

Winter Warriors

Not all of them leave, though. Hardy species remain in temperate forests through the coldest months, hunting for food wherever they can find it.

Winter in temperate forests can be desolate, and food is much harder to find. Many trees drop their leaves and plants go dormant to survive cold temperatures and a lack of sunlight. Insects hide themselves away. Lakes freeze over, putting fish and aquatic vegetation out of reach. The majority of birds found in temperate forests must depart in the winter—migrate—in order to find the food they need to survive.

Though insects hide away in the winter, woodpeckers are equipped to find them. Chisel-like beaks and shock-absorbing bones in their heads help woodpeckers bash and chip away at trees to find wood-boring insect larvae inside. Pileated woodpeckers, the largest species in North America, have a year-round penchant for carpenter ants, and may be seen in the middle of winter hacking apart a tree stump in search of hibernating ant colonies inside. Dozens of nonmigratory woodpeckers thrive in temperate forests by finding insects when other birds can't.

TEMPERATE FORESTS

They aren't the only birds eating meat in the winter: There are raptors around. While many raptors are migratory, other species stay all year to hunt mammals or birds. Some species are specially adapted to help them hunt in the most challenging conditions. The great gray owl's ears, like those of most owls, are positioned unevenly on their heads. Sound reaches each ear at slightly different times, helping the bird pinpoint the location of the noise, even the faint rustling of a rodent moving around under the snow.

> Silence helps owls listen for prey and sneak up undetected. To help keep quiet, they've evolved comblike structures on their wing feathers to dampen the "whooshing" noise other birds make when they fly.

Most other species sticking it out through a winter in a temperate forest eat plants. Seeds are the most abundant food source, and many species are equipped to eat them. Finches have a variety of bill shapes that help them access their particular favorites, from the sharp bills of siskins that help them pluck tiny seeds from alders and birches to the uniquely shaped, overlapping mandibles of the crossbills, which are used to pry apart tightly shut conifer cones.

Chickadees eat whatever they can find in winter, including any hibernating insects they stumble across, but they also prepare themselves for the cold days ahead by caching seeds. In the fall when seeds are plentiful, some birds will tuck seeds into nooks and crannies in tree bark to save them for when times are tougher. Chickadee brains are known to expand in winter as they work to remember their network of hidden snacks. Tits and nuthatches are also known to engage in caching, as well as larger birds like jays, crows, and nutcrackers.

The winter is even more challenging for birds that live on the ground. A blanket of snow makes walking difficult for ruffed grouse, but it, too, has evolved to cope with life in the winter. In autumn, it grows small fingerlike appendages called "pectinations" on the sides of its toes. These growths give the grouse's foot a wider surface area, acting like a kind of snowshoe and helping the bird walk on top of the snow without falling through it.

Our impact on birds in the temperate forest affects winter species and migrants alike. Human development has resulted in the loss of millions of acres of habitat for birds and other wildlife. Centuries of our occupation of temperate forests has led to overhunting, physical changes to the rivers and wetlands, and air and water pollution. And in just the last few decades these forests are beginning to feel the impacts of human-caused climate change.

In some cases human actions have resulted in the extinction of bird species in the temperate forest, like the passenger pigeon, which was once the most populous bird on the entire planet. More recently, we humans have caught ourselves before it's too late, and working to change our behavior to protect other species. The bald eagle was once nearly extinct due to the unregulated use of dangerous pesticides but has now fully recovered its population in just a few decades.

TEMPERATE FORESTS

Still, there are far fewer birds in our temperate forests than there were before humans came along—billions and billions fewer. The populations of hundreds of species continue to fall. It's worth imagining what these forests might be like if all the birds disappeared.

IN A WORLD WITHOUT BIRDS: SEED DISPERSAL AND POLLINATION

Perhaps nowhere on Earth would the loss of birds be as devastating as to the kingdom of plants. Birds are essential to the lives of many plant species, either by protecting them from insects, by pollinating their flowers, or by transporting their seeds. If birds were gone, thousands of plant species could follow.

The loss of birds would desolate temperate forests and the rest of the planet by denying plants some of their most important pollinators. Flowering plants need to get pollen from one plant to another in order to reproduce, but they can't do it on their own. So, they entice animals to help. Plants produce flowers with edible nectar, and butterflies, moths, reptiles, mammals, and birds come to take a drink. Bits of pollen are transferred to the animal when they drink, and then they're carried to other flowers as the animal continues to feed, completing pollination.

WORLD WITHOUT BIRDS

Thousands of bird species aid in pollination in habitats all over the world. Sunbirds pollinate flowers across Africa and Asia. In the deserts of South Australia, brown honeyeaters help pollinate the iconic Sturt's desert pea. Honeycreepers are the most important pollinators of several plants on the Hawaiian Islands, and honeyeaters, spiderhunters, parrots, and many other birds pollinate flowers throughout the tropics.

But perhaps the most famous of all pollinators is the hummingbird. These tiny birds—the smallest of any on Earth—are found throughout North and South America in the appropriate seasons and have evolved bodies specially designed to sip nectar from and pollinate flowers. They're famous for their ability to hover in place on wings that can beat between ten and fifteen times per second, an adaptation that permits the birds to sip hanging flowers without having to perch. Their bills have become long and thin in order to help them reach deep into the nectar reserves of flowers, and in some cases have evolved to become curved or extra long to help them match the shape of flowers from specific species. The hummingbird and the flowers are in it together. Hummingbirds

Black-chinned Hummingbird

have evolved to drink nectar, and plants have evolved to encourage hummingbirds to visit. Flowers can exhibit certain features that are more attractive to birds than bees or other pollinators, including nectar, brightly colored flowers, and deeply placed pollen, while losing scent and landing platforms. In all, scientists estimate that around 7,000 plant species are specially evolved to favor hummingbirds.

The sharply curved bill of the white-tipped sicklebill is designed to fit only a few species of flower, such as those of *Centropogon granulosus*. The sword-billed hummingbird of the Andes has the longest beak in relation to its body size of any bird on Earth and is one of the only pollinators for the flowers of *Passiflora mixta*. With all of these adaptations, hummingbirds are the most common avian pollinators in North and South America, but but avian pollination is found all around the world.

So, if birds disappeared, a lot of plants would lose their pollinators. Thousands of species of flowering plants that have been involved for millennia in relationships with certain species of hummingbird, that changed their whole appearance to attract hummingbirds over other potential pollinators, would be left high and dry. Cardinal flowers, trumpet vine, and other forest flowers pollinated by the ruby-throated hummingbird; the Sturt's desert pea in Australia; bellflowers and other Hawaiian flowers pollinated by honeycreepers; and thousands of tropical plants around the globe would suddenly have a much harder time reproducing. Many of those plants would go extinct. Others might survive if insects picked up the slack. Either way, the loss of birds would cause an immediate and devastating impact to flowering plants around the world.

But the birds' usefulness to plants doesn't stop at pollination: They are also critical for dispersing seeds. Plants can't get up and walk around, so they need help if they want to grow in new areas. Some plants, like certain oaks or dandelion flowers, have evolved to produce seeds that can be carried by the wind. Other seeds, like coconuts, move by floating on water currents. Plants also travel by tricking birds into flying their seeds around.

The trick is that plants get birds to eat berries with seeds in the center. After flowering, many trees and shrubs in the temperate forests produce berries: tasty, brightly colored foods that are often the perfect size for a bird to swallow whole. Berries serve no purpose other than to entice an animal, most often a bird, to eat them. The bird eats the fruit but the seed at the center stays intact, and hangs out in the bird's digestive system while it flies around. When the bird poops, the seed falls to the ground and takes root in its new home.

Seed dispersal from birds is essential to the survival of some ecosystems, especially those with few other large, fruit-eating animals like monkeys. The plants have evolved their fruit to optimize its enticement to birds and other frugivores, using particular colors and scents to encourage their fruits to be eaten and their seeds spread. In fact, more than 70 percent of flower-bearing plants rely on birds to help disperse their seeds.

Droppings aren't the only way plants use birds to get their seeds around. Another plant trick is to get birds to carry seeds by accident. Various plants produce seeds that stick to all kinds of things, like the hair of a mammal or even a pair of hiking boots, using hooks or barbs or sticky substances. When the bird preens the seeds off its body, they drop to the ground and begin to grow. Sticking seeds are a feature of many plants in temperate forests, like burdock, mistletoe, tick trefoil, and many more.

Birds also transport seeds when they eat them. Instead of soft fruits like berries, some plants produce hard fruits called nuts, which are also eaten by birds. But the hard shells of nuts make them last longer than berries and other soft fruits, which can rot quickly, and so birds and other animals have learned to gather nuts and seeds and save them for later. Crows and other birds in the corvid family are famous for hiding "caches" of nuts around their territories in the fall to prepare for winter when food may be difficult to find. Hidden away underground or in another out-of-the-way spot, the nuts and seeds have an opportunity to germinate.

Plants that produce fruit and berries would be decimated by the loss of birds. Fruit and nuts would fall to the ground in piles and rot, uneaten, and wouldn't grow in the shade of their parents. Or they could be eaten by insects or other animals that could either digest or break up the seeds, preventing them from growing. Without the ability to spread into new areas, their populations would crash and be taken over by plants with other reproductive strategies.

In some parts of the world we don't need to speculate about what the loss of seed-dispersing birds would do. In the 1940s humans accidentally introduced the brown tree snake onto the small South Pacific island of Guam, where it promptly began climbing trees to eat eggs and nesting birds. Populations of native birds and other of the island's forest animals collapsed. All the seed-dispersing birds on the island were gone, leaving only fruit bats to help spread seeds. In the absence of the birds, scientists found that just 10 percent of a tree's seeds were now making it beyond the immediate area of the parent tree, compared to 60 percent of the seeds on other islands with birds. Overall, the rate of new tree growth in Guam dropped by 92 percent. Hundreds of plant species in the temperate forests that rely on birds to disperse their seeds, and thousands more plants around the world, would face potential extinction without birds to help reproduce and move them.

OVERHUNTING AND THE EXTINCTION OF THE PASSENGER PIGEON

Humans need to eat, and birds have always been on the menu. From ancient civilizations to modern times, hunting birds has been both a sport and a means of sustenance. But we must be careful. Unregulated hunting can overwhelm bird populations and cause entire species to go extinct.

The passenger pigeon was once the most populous bird on Earth. Scientists estimate that there were somewhere between 3 and 5 *billion* passenger pigeons in North America around the time Europeans arrived in the 1500s, meaning there were somewhere between six and ten times as many passenger pigeons in America as humans in the entire world. By 1914, they were completely extinct.

Passenger Pigeon

How could it be that a bird that was estimated to have made up somewhere between 20 and 40 percent of all bird life in America was wiped out in just a few hundred years? If the passenger pigeon could become extinct, is there any species numerous enough to survive? How was it possible for this to happen?

The passenger pigeon was a medium dove approximately the same size and shape as the common mourning dove. Adult males were gray backed with a wine-colored neck and breast, while adult females were brownish overall. They were birds of the deciduous forest—their preferred food was nuts or seeds of hardwood trees like beech and oak, but they also ate fruits and berries, insects and snails, and agricultural grains—and ranged across the eastern United States.

Early Americans shot passenger pigeons for food, and they made contests out of shooting them—just blasting guns into the massive overhead rivers of pigeons—for fun. European settlers in the Americas hunted passenger pigeons with a mindless, giddy, unquenchable abandon.

French explorer René Laudonnière wrote of killing close to 10,000 birds around Fort Caroline, Florida, in a matter of weeks to feed his soldiers. Methods of killing became more efficient over the years. Competitions sprung up where shooters would spread out underneath a flock and try to kill the most birds. In one competition, the winner killed more than 30,000 pigeons. Huge, tunnel-shaped nets were constructed, trapping as many as 3,500 pigeons at a time for sale by the barrelful. Hunters burned sulfur beneath the birds to suffocate them out of trees, or they cut down the branches they nested on.

People started noticing the decline of the passenger pigeon as early as the 1850s, and some states made attempts to pass laws preventing widespread slaughter in the coming decades. But they were either all too late or impossible to enforce, and habitat loss also contributed to dwindling populations. The last known passenger pigeon, a bird named Martha, died in the Cincinnati Zoo in 1914.

To those early Americans describing the infinite swarms of passenger pigeons, the thought that such a species—of all species—could completely disappear may have been as unbelievable as the sight of the flocks themselves. The extinction of the passenger pigeon served as a wake-up call to humanity: No species, not even the most plentiful on Earth, is immune to our concentrated destruction.

PESTICIDES AND THE AMAZING RECOVERY OF THE BALD EAGLE

At fourteen pounds and with an eight-foot wingspan, the bald eagle is one of the largest raptors in North America. An apex predator, the bald eagle occupies a broad range

Bald Eagle

of environments, inhabiting river, coastal, and lakeside habitats from tropical Florida to the farthest of the Aleutian Islands in Alaska. There were an estimated 100,000 pairs of bald eagles in North America before European colonization.

However, beginning in the late 1940s, ornithologists studying bald eagles noticed sharp declines in their populations. Something was destroying their ability to reproduce. Eagles were either not producing any eggs at all, or the eggshells were so thin and brittle that eagles would crush their own eggs while incubating them.

One researcher and writer, Rachel Carson, began putting the pieces together. She noticed that areas experiencing sharp bald eagle declines had all been sprayed with insecticides. She published her discoveries about the destruction caused by rampant use of chemical pesticides in a book called *Silent Spring*, one of the most influential works of environmentalism ever produced.

In *Silent Spring*, Rachel Carson focused on one of the first synthetic insecticides, a chemical called *dichloro-diphenyl-trichloroethane*, or *DDT*. The chemical was sprayed to kill insects in massive quantities: over farm fields to protect crops, over forests to kill timber parasites, and over towns and cities to control pest insect outbreaks.

But DDT was not a targeted poison—it didn't just kill the target pest insects but almost all others. The chemical kills by disrupting the nervous systems of insects, causing spasms and death. Worse, DDT was stored in the bodies of creatures that eat insects, a phenomenon called *bioaccumulation*. Carson found that DDT was being sprayed to control insects, and the chemical was moving up through the food chain from insects to fish to eagles.

When Rachel Carson published *Silent Spring* in 1962, it was a national sensation. Chemical companies spent hundreds of thousands of dollars in efforts to stop publication of the book and discredit its factual claims. They called her a "Communist" and a "food faddist" and said she was emotional, radical, unpatriotic, and "hysterically overemphatic." But they couldn't stop the truth. President John F. Kennedy convened a commission to study pesticide use, and it vindicated Carson and her claims. The newly formed Environmental Protection Agency banned the use of DDT in 1972, and the bald eagle was listed under the Endangered Species Act soon after it was passed in 1973.

Then bald eagles began to recover. With DDT out of use and the government working to reintroduce eagles and protect their habitat, eagle numbers climbed. Other birds of prey impacted by DDT experienced similar increases, like the peregrine falcon and osprey. Incredibly, the population of bald eagles in the lower forty-eight states increased from nearly zero in 1970 to more than 316,000 in 2021. It is, to quote Interior Secretary Deb Haaland, a "historic conservation success story."

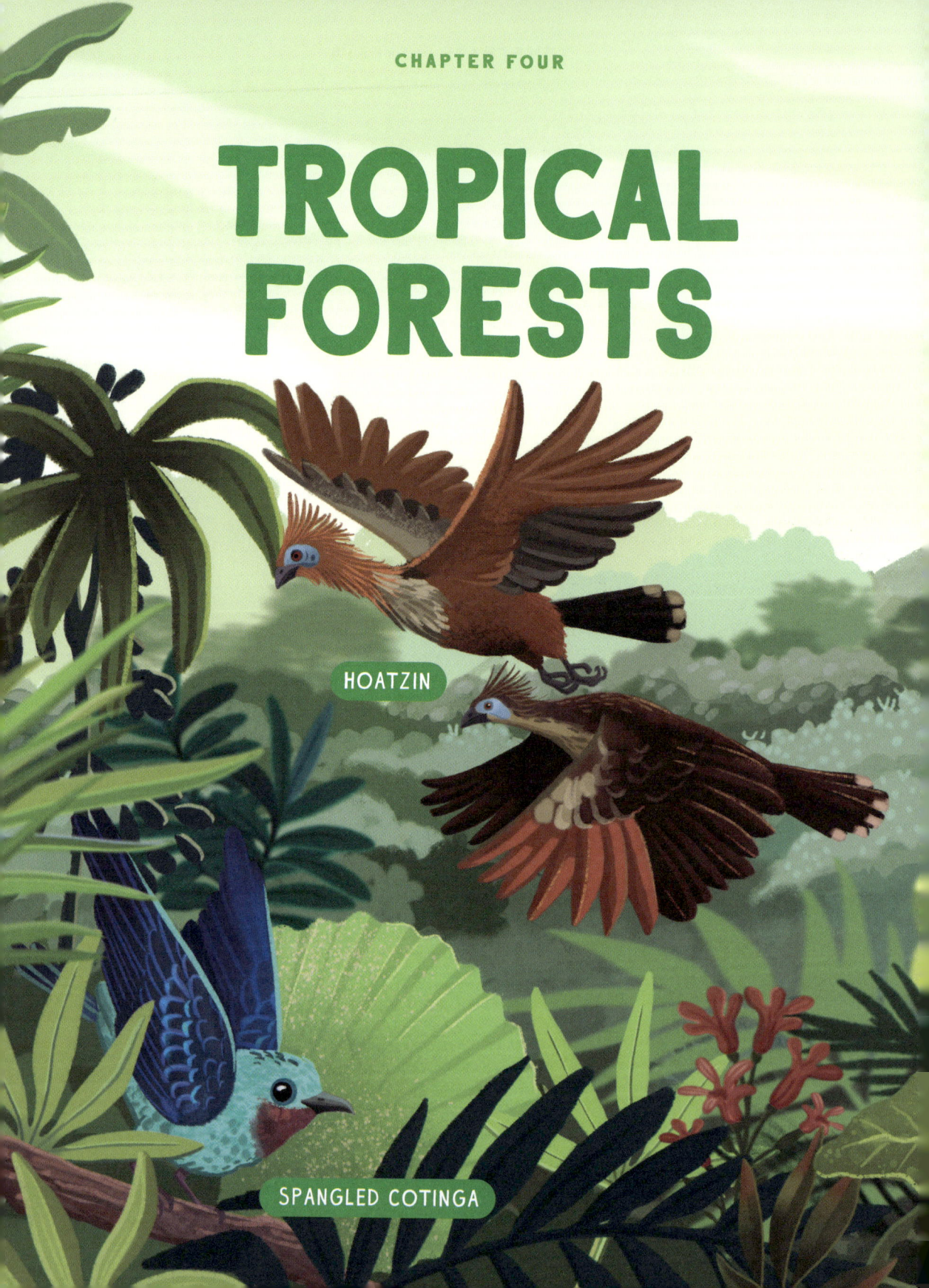

WORLD WITHOUT BIRDS

Unlike the temperate forests, which experience changing seasons throughout the year, the tropics have a warm, steady climate all year long. High temperatures, plenty of sunshine, and lots of rain mean that tropical forests are some of the most biologically rich areas on the planet. Though they cover less than 10 percent of the surface of the Earth, tropical forests contain about two-thirds of our biodiversity—including thousands of species of birds.

Toucans, sunbirds, hummingbirds, parrots, birds of paradise, tanagers, and many other gaudy species liven up tropical forests that ring the globe on either side of the equator. Central America and northern South America, central Africa, Southeast Asia, northern Australia, and many islands in between all have tropical forests, which may also be called jungles or rainforests.

Year-round growing seasons mean that plant life thrives. Tropical plants compete with one another to soak up precious sunlight, forming dense canopies of leaves that capture up to 98 percent of sunlight before it reaches the forest floor. The bounty of plant life in tropical forests means plenty of food for other animals. Leaves, seeds, fruit, and nectar are all eaten by a variety of herbivorous animals, who are then preyed upon by carnivores. There are birds that play each of the roles the rainforest provides: fruit-eater, insect-eater, apex predator, and more.

TROPICAL FORESTS

Tropical forests are more than just wildlife hotspots, they're also globally important ecosystems that store billions of tons of carbon, pump oxygen into the atmosphere, and produce and filter rainwater. Yet tropical forests are under severe threat. Scientists have found that of the planet's more than 5.6 million square miles of tropical rainforest, 34 percent is completely gone, 30 percent is polluted or fragmented, and just the remaining 36 percent is intact.

Birds Everywhere

Tropical forests are so dense that distinct layers exist at different heights of the forests. Some species find everything they need at the tops of trees, for example, and never have any reason to head down to the forest floor. Tropical forests are often thought of as having four distinct layers: the very tops of the trees, called the *emergent layer*; the rooflike tangle of branches and leaves of the *canopy layer*; the crowded and brushy *understory*; and the dark and leaf-strewn *forest floor*.

Each level has its own specialty birds. Large birds of prey nest in and patrol the emergent layer, sometimes more than 200 feet above the ground, where few other predators can maneuver. The harpy eagle of Central and South America is an apex predator, possessing the largest talons on Earth and taking on large prey. Weighing up to twenty pounds each, the eagles scan through the upper branches looking for food and then swoop silently down onto their target. Sloths and monkeys are their favorite prey, but they also eat large parrots, iguanas, squirrels, coatis, and basically anything else they can carry.

The canopy layer hosts more wildlife than anywhere else in the tropical forest. Here, overlapping branches form elevated pathways for mammals to move from treetop to treetop, and present plenty of places for birds to perch. The trees and other plants provide fruit and nuts for wildlife to eat, and birds feast. Some of the most famous are the hornbills, a group of tropical birds with huge, down-curved bills. Hornbills travel in pairs or small flocks through the forests of Asia and Africa, looking mainly for ripe fruit. A hornbill's bill is so long that its tongue can't reach the end, meaning it needs to toss fruits from the end of its bill into its mouth with a flick of its head.

Below the canopy is the understory. Here, much of the sunlight is blocked by the leaves above, and young trees and shrubs fight each other for what light makes it through. It is a dense and humid place, ideal for insects—and for the birds that eat

> Depending on the species, the structure on a hornbill's bill, called a *casque*, may be used to strengthen the bill, act as a vibrating chamber to make the bird's calls louder, or serve as a weapon for aerial combat.

TROPICAL FORESTS

them. Some birds found in the understory aren't permanent residents. Tropical forests around the world host billions of birds that spend their summers in the temperate forests and then fly back to the tropics in winter to feast on the insects that live there year-round. Warblers,

Yellow Warbler and Alder Flycatcher

flycatchers, vireos, cuckoos, and many more birds can be found deep in a tropical forest in winter, and nesting in a suburban backyard in summer.

The final layer is the forest floor. Birds here search for insects and other prey underneath the leaf litter, or scavenge for fruits and nuts fallen from above. Some of these species spend all their time on the ground, like the massive southern cassowary of Australia and Papua New Guinea, or the red jungle fowl of Southeast Asia—the ancestor of our modern chicken. Other birds are just visitors, including birds that follow ant swarms in tropical forests around the globe. Some tropical ants form massive raid parties, marching along the forest floor eating anything they can capture. Dozens of bird species follow, including more than eighteen species of antbirds in Central and South America, not to eat the ants but instead to snag creatures scurrying away. There's food for every bird somewhere in the tropical forest.

Shining Bright

For some tropical birds, food is so abundant that they can worry about other things—like how to impress a mate. Tropical birds are famous for being more flamboyant and colorful than bird species in the temperate areas to the north and south—think of the long yellow beaks of toucans, the gaudy rainbow plumage of parrots, and the striking throat patterns of hummingbirds. Studies have shown that not only do tropical birds come in a broader spectrum of colors than temperate birds, but their colors are more vibrant.

Scientists have a few theories as to why this is. One is that birds living in dense, dark forests need to be bright in order to stand out. Another points out that tropical birds tend to eat more fruit than birds elsewhere, and that the colorful pigments in the fruits influence the birds' plumage. Perhaps the most likely theory is that birds in the tropics have more energy with which to produce colorful feathers. Bright colors require more resources to produce than duller ones, and many species in temperate climates need to put their energy into finding food or keeping themselves warm or flying away from predators. Tropical birds don't need to keep themselves warm because it's always warm in the tropics, and there is often such an abundance of fruit and insects that birds don't need to spend much effort feeding themselves. They've got extra energy to put into looking snazzy.

And looking good helps to attract mates. For most bird species in the tropics—and around the world—females choose who they want to mate with. They want birds that are strong and healthy, and sporting a plumage with the brightest colors is a sure way to let a female bird know that you're fit.

For some tropical species, just looking good isn't enough. When there is plenty of food to go around and all the males are looking sharp, they may need to do more to differentiate themselves for females: They need to perform. Certain tropical birds have evolved incredibly elaborate feathers—well beyond just bright colors—and lively dance routines in order to attract mates.

Take the manakins of Central and South America, for example. Many species of these small songbirds use horizontal branches as dance floors to perform fancy routines, including wild stutter-steps and moonwalks, synchronized flight displays, twirls, head bobs, and other moves. Some, like the club-winged manakin, have evolved special modified feathers on their wings, which they shake rapidly to produce sounds like sirens, pops, and whooshes. All to impress females.

> The tail of the ribbon-tailed astrapia, a species of bird of paradise found in Papua New Guinea, can be more than five feet long. At over four times the length of the bird's body, it is the longest tail in relation to body size of any bird on Earth.

The most extravagant of all tropical birds are the birds of paradise found in Indonesia, Papua New Guinea, and Australia. Male birds of paradise have evolved incredible plumage in order to interest females, including head plumes, extra-long tail feathers, curly tail feathers, ruffles, wiry feathers, and almost everything in between, all in bright, bold colors. Males strut their finery on carefully selected stages, each meticulously cleared of leaves and other debris, executing dramatic and mesmerizing dance moves for potential partners. The feathers and the dance moves all take energy and effort that birds outside the tropics cannot spare. However, the abundant food found in tropical forests leaves birds with enough free time to party.

IN A WORLD WITHOUT BIRDS: INSECT APOCALYPSE

If birds disappeared, a silence would fall upon tropical forests previously alive with birdsong, but it wouldn't last long: Soon there would be other sounds. Buzzing. Whining. Clicking. Chirping. The sounds of insects.

Raggiana Bird-of-Paradise

Without birds around, insects and other bugs would proliferate. The forests would teem—even more than they do already—with mosquitoes, beetles, flies, ticks, ants, spiders, bees, moths, butterflies, and all kinds of other invertebrates.

Birds eat a *lot* of insects. One study determined that the world's insectivorous birds eat between 450 and 550 million tons of insects each year. 550 million tons! That's almost twice the weight of all the humans on Earth! That's 20 *quadrillion* individual bugs. I didn't even know there was such a thing as a quadrillion!

The same study found that the biome with the largest share of these bugs—some 217 million tons of them each year—was the tropical forests.

Have you ever been to a place with lots of bugs? Someplace where you were constantly being bombarded with mosquitoes or flies, or maybe there were a ton of ants crawling around? Okay, now picture yourself in that same place but with 20 *quadrillion* more bugs around. That's what it would be like if birds disappeared.

The impact to your camping trip would be bad, but the impact to the rest of the forest would be devastating. Insects and bugs are some of the main predators of plants, and if birds disappeared, insects would multiply unchallenged and eat their way through the forest. Caterpillars, beetles, ants, and other critters would strip leaves bare, leaving defoliated skeletons in their wake.

In certain parts of the world, the impacts of massive insect populations can already be seen. Sometimes, native insects go through population booms that overwhelm the ability of native birds to control. Outbreaks of eastern spruce budworm populations every few decades kill millions of spruce and fir trees. Other times, humans accidentally transport insects from one part of the world to another, where they may not have any natural predators. These insects can run rampant in their new home, eating their way through a forest with little resistance. The emerald ash borer was accidentally brought from its native range in East Asia to the United States, where it has killed more than 100 million ash trees in just decades.

Humans would suffer. Insects already eat a lot of the crops that we plant for our own food—somewhere between 5 and 20 percent of all grain crops. In the tropics, an estimated 40 percent of the global cocoa crop is lost to pests and disease, along with billions of dollars more to insects eating guava, coffee, papaya, and other tropical crops. Without population control from birds, these losses would mount.

The human response to the increased agricultural threat may be even worse. Chemicals used to control insects and protect crops—pesticides—are already a major source of environmental pollution. Excessive use and misuse of pesticides contaminates soil and water, impacting not just birds but all life. Humans would surely ramp up their use of insecticides to protect crops, putting more pollution onto the landscape and further reducing biodiversity.

TROPICAL FORESTS

There would be even more direct impacts to humans. Insect- and arachnid-borne diseases, most often transmitted by mosquitoes, ticks, tsetse flies, and fleas, affect millions of people each year and kill hundreds of thousands. More than 200 million people contract malaria each year as a result of mosquito bites, and an estimated 400,000 die. Dengue, chikungunya fever, Zika virus, yellow fever, West Nile fever, Japanese encephalitis, tick-borne encephalitis, bubonic plague, sleeping sickness, American trypanosomiasis, and other diseases have all made their way to people via insects and bugs.

The other wildlife in the tropical forests would suffer, too. Those that eat vegetation—deer, primates, squirrels, sloths, and more—would lose food to the ravaging insects. They'd also be beset by pests themselves: blood-sucking ticks and mosquitoes, and biting flies.

There would be few beneficiaries of a major increase in insects and other bugs in a world without birds. Bats, certainly, would find themselves with more food to eat. Without birds to eat caterpillars, there would be many more night-flying moths for bats to pick off, and bats may find themselves with more roosting spots available with fewer birds occupying tree cavities and other tight spaces. Other insect-eating creatures, like spiders, opossums, skunks, shrews, sugar gliders, echidnas, moles, and more, would also likely benefit from the absence of birds.

But could they increase enough to eat 20 quadrillion more bugs? Who knows? It's a number beyond comprehension (apparently a quadrillion is a thousand trillions). Without birds to eat them, those insects would swarm through the tropics and around the world.

CLIMATE CHANGE AND THE EXTINCTION OF THE PO'OULI

Human activities are the driving force behind the alarming changes in the Earth's climate. Through the combustion of fossil fuels, deforestation, industrial processes, and various other practices, we have significantly increased greenhouse gas emissions, leading to a measurable rise in global temperatures. Changing temperatures have far-reaching consequences, such as the melting of polar ice caps, rising sea levels, altered weather patterns, more frequent and intense natural disasters, and disruptions to ecosystems and biodiversity.

When climate change influences a habitat, new species can move in, including deadly ones.

Po'ouli

Humans have introduced many nonnative species to Hawaii, including cats, dogs, rats, and pigs. But of all the introduced species in Hawaii, the mosquito may be the most dangerous. Around the world, mosquitoes carrying diseases like malaria, dengue, West Nile, yellow fever, Zika, chikungunya, and lymphatic filariasis kill hundreds of thousands of humans each year, and they kill native Hawaiian birds as well. The introduced house mosquito (*Culex quinquefasciatus*) carries avian malaria and avian poxvirus, two diseases to which native Hawaiian birds have little resistance. Defenseless, many birds succumb to the disease.

Mosquitoes thrive in warm weather and had for decades been absent from the higher-elevation tropical forests on the Hawaiian Islands, giving native birds some respite from their attacks. However, mosquitoes fly higher as the climate warms, squeezing native species into smaller and smaller ranges. One such species, the po'ouli, only became known to ornithologists in 1973.

The po'ouli was understood to be imperiled almost as soon as it was found. With its range limited to the fragment of forest it was discovered in, the entire po'ouli population was estimated to be no more than 200. Conservation efforts came too late. In the 1990s, extensive surveys turned up just three living po'oulis.

In 2004, scientists decided to try to capture all three birds and attempt to breed them in captivity, but only one could be found. That bird, a very old male, was captured but died before the year was out.

No po'oulis have been seen alive since. The species was officially declared extinct in 2021, less than fifty years after it was discovered.

Scientists are still working across the islands to save the remaining seventeen species of Hawaiian honeycreeper. Mosquito-borne illness remains the greatest threat, but it's difficult to control a millions-strong mosquito population. The use of insecticides could harm other wildlife. Some argue for the release of sterile, genetically modified mosquitoes, hoping that they will mate with existing mosquitoes but not produce offspring. Alternatively, mosquitoes born in a lab from eggs infected with a particular strain of native bacteria are unable to reproduce, and can result in fewer mosquitoes. These methods are costly, work intensive, and not guaranteed to succeed, but there are few other options.

The tropical forests on Hawaii are just a fraction of the size they once were, and a warming climate is allowing mosquitoes and other invasive species to creep into some of the forests' last havens for native species.

TROPICAL FORESTS

THE PET TRADE AND THE SURVIVAL OF THE HYACINTH MACAW

Hyacinth Macaw

Sometimes we hurt birds by loving them too much.

Humans have been keeping pets for thousands of years. Parrots might be the most popular pet bird of all. These long-lived, beautiful, and intelligent birds have been companions for the ancient Egyptians, Greeks, and Romans; Indigenous Americans; kings, queens, popes, and emperors in Europe; and United States presidents. There are estimated to be more than 20 million pet birds in the United States, many of them parrots.

As beloved as pet birds are, the hobby is controversial when birds are taken from the wild. The international wildlife trade is a multi-billion-dollar industry that has involved both the legal and illegal exchange of millions of birds. At least 2,345 bird species—more than a quarter of all known—have been recorded in the wildlife trade.

Parrots are by far the most frequently stolen bird, making up nearly 90 percent of all live birds traded. Nearly half of all parrot species are considered either endangered or threatened across their worldwide tropical range. But for some, like the hyacinth macaw, there's hope.

The hyacinth macaw is a showstopper. When its striking cobalt wings and tail are extended, they reach a full three feet, longer than any other parrot in the world. Its entire body is bright blue save for bright yellow spots of bare skin on its face. Its bill is huge and black and hooked, and strong enough to crack open tough palm nuts.

Hyacinth macaws are among the most prized parrots for collectors of exotic pets, and the demand has led to many parrots being illegally removed from the wild. Smugglers climb nest trees to snatch eggs, or to lure adult birds into sticky lime traps.

By way of these methods and others, an estimated 10,000 hyacinth macaws were taken from the wild in the 1980s and sold on the international black market for as much as $12,000 apiece. Habitat loss and human hunting for food were also taking a toll across the species' disparate range. At the end of the decade there were only estimated to be around 2,500 birds left in the wild.

In the late 1980s a young biology student named Neiva Guedes watched hyacinth macaws flying over the Pantanal, a region of marshlands in South America, and grew alarmed at news that they might soon be extinct. She founded the Hyacinth Macaw Project in 1990 and works with ranchers and locals to protect preferred nesting trees and install nest boxes. She wanted ranchers to feel proud that

TROPICAL FORESTS

they had nests of these beautiful birds on their land. With greater awareness of the bird and its needs, there'd be more incentive to prevent poaching and protect habitat.

Hyacinth Macaw

It's working. The number of hyacinth macaws in the million-acre area of the Pantanal covered by the Project has more than tripled since her project began. Conservation—even when conducted by small, underfunded teams—can work to reverse declines and save species.

Unfortunately, not all populations of hyacinth macaws are faring well. Though the illegal parrot trade for this species has slowed, habitat loss from expanding cattle ranches and agriculture has kept the species under pressure. Numbers continue to fall in the macaw's other ranges outside the Pantanal, but there is hope. Perhaps the locally driven, hands-on success of the Hyacinth Macaw Project will allow populations to recover, and that its approach could work in other parts of Brazil.

WORLD WITHOUT BIRDS

Life is tough at the poles. **The farthest northern and southern portions of our globe are cold and inhospitable places, frozen and lonely. The continent of Antarctica is almost entirely covered with ice, with an average temperature during the summer—if you can even call it that—of -18°F, and winter temps averaging a deadly -78°F. The North Pole is comparatively balmy, averaging right around the freezing mark of 32°F during its short summers.**

Yet birds live even here, in the harshest of environments. In fact, the seas at the edge of the Arctic ice and along the coast of Antarctica are positively teeming with bird life. Millions of seabirds journey to newly ice-free Arctic seas each summer to breed. The Antarctic Ocean is similarly productive, with millions more seabirds nesting on remote islands off Antarctica and massive penguin colonies crowding the continent's coast.

Equally important to birds is the Arctic tundra. The northern latitudes of North America, Europe, and Asia are too cold and the growing seasons too short to permit the growth of trees. Instead, these areas are vast expanses of short grass, moss, and shrubs, spotted with lakes and ponds. The short summers and brutal winters keep overall diversity low, but the remote and relatively safe tundra is a refuge for millions of migratory birds who show up for the summer and leave before the ice returns. In all, at least 200 species of birds breed on the Arctic tundra in summer.

However, the rich bounties of the polar regions have long attracted humans as well. Whalers were some of the earliest visitors to many isolated southern islands, and their presence forever altered many of the ecosystems through the introduction of livestock as a stable food source; the accidental introduction of mice, rats, and other pests; and the direct killing of nesting seabirds for food.

Though the polar regions are remote and severe, their birds are still at risk.

Palace of the Penguins

The southern continent of Antarctica is the harshest place on Earth. The land itself is covered in snow and ice year-round, and it is the driest, coldest, and windiest place anywhere. There's nothing to eat on land except for some mosses and lichens and a few plants that hang on in exposed areas. Just one native insect, the Antarctic midge, is found on the mainland of the continent. There are no native

Emperor Penguins

land mammals, and barely any life at all near the immediate coast.

The surrounding seas and nearby islands, though, are full of life. The Southern Ocean, also known as the Antarctic Ocean, hosts abundant marine life, and if you want to survive here, you better be ready to swim for your supper. Fish and krill are dined on by a host of predators including seals, whales, and some of the most accomplished swimmers in the Southern Ocean: penguins.

There are eighteen species of penguin in the world, most of them living in the southern hemisphere. They evolved from a common ancestor related to albatross and petrels—birds that already spend a lot of time on the water catching fish. Many of today's flying seabirds spend significant time underwater, using their wings to propel themselves after fish. At some point, the ancestors of penguins spent so much time underwater that their wings became better used for swimming than flying, and they lost the ability to fly altogether.

They became incredible swimmers. Penguins are torpedoes underwater, moving quickly and with great agility after their prey. They're amazing divers, and the emperor penguin can dive down to

Gentoo Penguin

THE POLES

1,800 feet below the surface, deeper than the dive record of dolphins and some other marine mammals.

However, like all birds, penguins must come to land to lay eggs. Coming ashore is a daunting proposal for penguins. Their bodies are evolved for swimming, so they're awkward on land, and their inability to fly means they're vulnerable out of the water. But they've chosen to live in safe areas: Antarctica and the islands in the Southern Ocean have no native mammalian predators on land to attack penguins the way a polar bear or wolf could in the Arctic, and so penguins can lay their eggs in safety.

> Most penguins live in the southern hemisphere, but one species, the Galapagos penguin, has managed to colonize the Galapagos Islands off the coast of Ecuador. They are the only penguin species that live north of the equator.

It's not easy, though, especially for those species that spend the winters on Antarctica itself. Just two penguin species, emperor and Adélie, live exclusively on the continent of Antarctica. Others breed on remote islands in the Southern Ocean, New Zealand, in southern Africa, and elsewhere.

The emperor penguin is the largest in the world, and its yearly breeding cycle is as extreme as the land it calls home. In March and April, at the beginning of the Antarctic winter, adult emperor penguins leave the sea and waddle their way up to seventy-five miles inland to nesting colonies. There females lay eggs and delicately transfer them to the feet of the males, who will keep them warm under their bodies—but off the ice—all winter long while the females return to the sea to feed. Winter temperatures regularly reach -40°F, and the male emperors huddle together to keep warm, taking turns at the cold edge and the warm center of the flock. Males hold the eggs for up to seventy-five days before the eggs hatch. Females return soon after hatching with food in their bellies, and the males are finally able to return to the sea to feed themselves.

On Top of the World

Life is a little different on the other side of the world, but it's still very cold. Unlike the South Pole, which is covered by the snowy landmass of Antarctica, the ocean at the North Pole is topped by a relatively thin layer of ice that expands and contracts throughout the year. Thus, the majority of wildlife in the Arctic is concentrated around the northern edges of large continents: North America, Asia, and Europe.

The comparatively large amount of land in the Arctic means there's a lot more room for birds to nest. In addition to rocky islands fit for seabirds, there are thousands of square miles of tundra: a vast, treeless landscape covered in snow and ice in the winter but bursting with mosses, lichens, small shrubs, and wildflowers and dotted with countless small ponds and wetlands during the brief summers. Millions of migrant birds pour onto the tundra each spring to breed.

Ducks, geese, and shorebirds find tundra wetlands, brimming with insects and aquatic arthropods, ideal places to raise their young. Small lakes lack the waves and tides of the ocean, but still provide protection from arctic foxes and other predators. The tundra is a perfect nursery, but would-be parents are on the clock as soon as they arrive. They must mate, lay eggs, and raise chicks until they can fend for themselves in the two or three months available before cold sets in again.

As in the Antarctic, the Arctic Ocean is a haven for life during the summer. Retreating ice and abundant daylight produce excellent conditions for phytoplankton and zooplankton, which in turn feed fish, seals, and whales. Seabirds feast, too, and many take the opportunity to breed. Unlike their Antarctic cousins, however, the presence of large mammals in the Arctic makes breeding a little more difficult. Arctic seabirds would be easy meals for polar bears, wolves, arctic foxes, and other predators if they nested right out in the open like seabirds do elsewhere, and so Arctic seabird nesting colonies are a little more extreme: often on the sides of sheer cliffs.

The top of the world is ringed by incredible cliffside seabird colonies, sometimes hosting hundreds of thousands of individual birds. Cliffs provide a measure of safety from terrestrial predators, but nesting hundreds of feet in the air on a narrow ledge of rock has other dangers. There's not a lot of room—there are only so many suitable places to nest on a cliff—so birds crowd into every available nook and cranny. Murres are known to fit up to seventy birds in a single rocky square meter, putting eggs at risk of being trampled or covered with dirt or guano. Consequently, many seabirds have evolved eggs that are weighted at one end and pointed at the other, a design that scientists believe makes them stronger and, because one end sticks up into the air, less likely to be covered in gunk.

Common Murres

All this breeding activity—millions of birds on the tundra and millions more on cliffs and remote, rocky islands—occurs during the brief Arctic summer. Birds are a crucial part of the thriving summer ecosystem. They serve as both predators and prey during the short months of warmth, feasting on insects and small mammals and themselves becoming food for arctic foxes, gyrfalcons, and wolves. Arctic birds like snow buntings and ptarmigans are essential summertime seed dispersers, and the droppings of millions of Arctic birds help to cycle nutrients back through the soil. Most birds are only in the Arctic for a brief time, but they make the most of it.

IN A WORLD WITHOUT BIRDS: BIRDS AS PREY

If birds disappeared from the polar regions, other creatures would go hungry. One of the most feared predators for penguins in Antarctica is the leopard seal, an eleven-foot carnivore with inch-long teeth. The seals hide at the edges of the ice, waiting to ambush penguins as they dive into the water. Penguins make up nearly half of a leopard seal's diet in some seasons, and their absence would be devastating.

Birds are eaten by many types of animals around the world, and living in a world without birds would mean that many of these creatures would be scrambling to find something else to eat. It would be particularly difficult for *avivores*, animals like that leopard seal that are specially adapted for eating birds or for which birds make up a large part of their diet.

Some of the most specialized bird-eating animals are birds themselves. It makes sense, really: Flying permits birds to escape most land-based predators, but not other birds. Two birds of prey, falcons and the *Accipiter* genus of hawks, are especially well-suited for hunting birds, but each group takes a different approach. Genus *Accipiter* is built to hunt birds through the forest. They have evolved short, broad wings and long tails to act as rudders to help them maneuver through trees and dense shrubs during ambushes, and long legs to grasp small songbirds out of midair. Falcons, on the other hand, don't bother with ambushes: They're speed demons. Narrow, pointed wings help them chase down birds in the open sky. In fact, the peregrine falcon can reach speeds of more than 200 miles per hour during some of its hunting dives, giving ducks and shorebirds little chance of escape.

But birds can't escape all mammals.

In Africa, two types of wildcats, the serval and the caracal, are especially effective at stalking and snagging birds—sometimes right out of the air. The cats use their stealth to sneak up as close as they can to birds resting on the ground before rushing ahead. The alerted birds beat their wings to get into the air, but the cats are right behind them, leaping up to more than ten feet in the air to snag the birds as they try to fly away.

Even some insects have developed a taste for birds, including two of the largest spiders on Earth. Though the behavior is apparently rare, the massive South American tarantula was given its name, the Goliath birdeater, after it was seen dining on a hummingbird. In Asia and Australia, the huge orb weaver spiders in the genus *Nephila* are often seen catching birds in their webs and wrapping them up for a meal.

Catching birds out of midair is difficult, though, and most terrestrial predators take an easier approach and feed on eggs or young birds still in the nest. Mammals able to access seabird breeding colonies, like arctic foxes in some parts of the far north, can gorge themselves on defenseless seabird chicks. If they can't reach the nests—and most of the time they can't because they're situated on cliff faces—the foxes wait at the bottom for young birds to fail when attempting to take their first flight.

In fact, bird nests—filled with protein-packed eggs or defenseless baby birds—are an important food for many animals. For birds across the globe, nest predation is the primary reason that nests fail and is often the most important factor in reproductive success.

Countless creatures worldwide raid bird nests. In addition to the previously mentioned foxes and cats, species of skunks, weasels, badgers, mice and rats, primates, squirrels, and many more mammals have been known to dine at nests. Even some creatures that are otherwise thought of as herbivores are sometimes known to eat baby birds or eggs, like deer, sheep, and cows. They're just too nutritious and too easy.

Reptiles are also a threat to nests, especially snakes. Among the dozens of species of snake known to eat bird eggs are those in the genus *Dasypeltis*, which are found throughout Africa and have evolved to eat nothing but bird eggs. A *Dasypeltis* snake has an extremely flexible jaw that allows it to swallow an egg in one piece and then use specially evolved spines extending from its vertebrae to break the shell inside its body, ensuring that not a drop of the egg will spill out.

It's difficult to know exactly how much the disappearance of birds would affect bird predators. Certainly those species that feed exclusively on birds or bird eggs, like *Dasypeltis* snakes, would face extinction. Other creatures that depend on birds for large parts of their diets, like leopard seals, or coastal arctic foxes during the summer, would need to replace birds with another food. Competition would increase for remaining food sources—fish and krill in the case of the leopard seal—or predators would expand their palates to include new foods. For example, after eradicating most of the native birds on the South Pacific island of Guam, the invasive brown tree snake began eating lizards, rats, mice, and domesticated birds and their eggs.

Different predators would have different reactions to the loss of such an important prey item, but all would suddenly be finding themselves having to fight to survive in a world without birds.

OVEREXPLOITATION AND THE EXTINCTION OF THE GREAT AUK

Great auks were forced into extinction around 1844. They disappeared during a time when technology had evolved more quickly than our understanding of environmental conservation. The excitement of exploration—of finding new worlds and new creatures within them—blinded us to recognizing that our impacts on those new worlds were immediate and maybe permanent.

But what an amazing sight a great auk colony must have been! Each spring, millions of these flightless birds would return to nesting rocks in the North Atlantic and scramble up through the surf and onto the rocks. At nearly three feet tall and eleven pounds,

Great Auk

great auks towered over their relatives in the Alcidae family that nested alongside them—Atlantic puffins, razorbills, common murres, and other diving seabirds.

Great auks and penguins aren't closely related, but the auks looked and acted similarly. Like penguins, great auks were slow and clumsy on land. They were also flightless, as their bodies had evolved to become agile underwater predators. By all accounts they were incredible swimmers. They used their small remnant wings to "fly" underwater, fast and sprightly in the pursuit of fish. They were also deep divers, reportedly holding their breaths for fifteen minutes at a time while they pursued prey hundreds of feet below the surface.

Early humans knew great auk colonies well. Archaeologists have found great auk bones at sites used by Neanderthals some 100,000 years ago, and the birds were hunted for their meat, eggs, and feathers by many other early peoples, including the Beothuk in Newfoundland, Inuit in Greenland, and Magdalenians along the coast of France and Spain.

Pressure on the great auk increased when Europeans began sailing across the Atlantic after they discovered North America. Great auk breeding islands were dotted all along the route between the continents and were perfect stop-off sites for sailors hungry for fresh food after long sea journeys.

Great auks, with their dense colonies, large size, and defenselessness, were a prime target. Their inability to fly not only made them easy to catch on land but also meant that their breeding islands were relatively accessible—since they couldn't fly up to cliffs or high rocks like other alcids, great auk breeding colonies generally featured gentler, sloping terrain that also permitted access to sailors. Such particular nesting-island requirements meant that only about ten islands fit the bill. Humans knew exactly where to find them.

Sailors decimated the colonies, but feather hunters upped the scale of destruction.

Certain North Atlantic seabirds at this time were prized for their down—the soft insulating layer of feathers beneath a bird's outer flight feathers. Down feathers from a particular species, the common eider, were collected in huge quantities for use in blankets, bedding, and clothing, but eiders were becoming very rare by the middle of the eighteenth century. Feather companies turned to great auks and further depleted the remaining populations.

In 1844, three Icelandic fishermen climbed onto Eldey, a tiny, rocky island that was home to the last known great auks, and found just a lone pair incubating a single egg. The men had been asked by a collector to obtain specimens, and they quickly killed the two birds and accidentally cracked their egg. The great auk was never seen alive again.

RODENTS AND THE INCREDIBLE RECOVERY OF THE SOUTH GEORGIA PIPIT

How do we fix the problems we've caused? With hard work, dedication, money, and, sometimes, with helicopters full of rat poison. That's what's happening on a remote, 100-mile-long island about 800 miles east of Argentina and 700 miles north of Antarctica called South Georgia Island.

This treeless and weather-lashed rock is about as inhospitable a place there is, but South Georgia Island is one of the most important

THE POLES

wildlife breeding colonies on Earth. The island hosts 30 million breeding birds, including 7 million penguins and 250,000 albatross, in addition to 2 million Antarctic fur seals and half of the world's population of southern elephant seals. It's also home to a small bird found nowhere else in the world: the South Georgia pipit. It's the only nesting songbird on the entire island and the southernmost songbird on the entire planet.

South Georgia Pipit

Human sailors first spotted South Georgia in 1765, and Captain Cook was the first to make landfall a decade later. He claimed it for England and named it after King George III. The island became an important hub for whaling ships and seal hunters in the subsequent decades, and tucked away into the holds of some of these early ships were the island's first mammals: rats and mice.

The native range of the brown rat (*Rattus norvegicus*) is believed to be China and Mongolia, but they're so widespread now that it's difficult to know for sure. Sometime in the Middle Ages brown rats began their human-aided march around the globe. Their inconspicuousness makes them good stowaways, and throughout their history they've been particularly adept at hiding on ships sailing around the world. Brown rats are large and hardy with an omnivorous diet—tough enough to survive and thrive in a variety of habitats.

Seabird colonies are particularly vulnerable, including on South Georgia. The rats nested in the tussock grass and fed on whatever they could find: grass, beetles, and nesting birds. By the early 2010s, the South Georgia pipit was listed as near threatened, and its numbers were plummeting. If something hadn't been done, the pipits and other seabirds of South Georgia may have gone extinct. But, incredibly, something *was* done. A huge something.

In 2011 a team led by the South Georgia Heritage Trust began planning the monumental task of eradicating the rats. Twenty-five people, including helicopter pilots, doctors, engineers, and three chefs, spent hundreds of hours flying over the island and, working against the changing seasons, poisoned the rats before birds returned to breed.

Only total eradication would do. If any rats were left on the island, they could quickly repopulate. The final phase of the plan, conducted in 2017, was perhaps the most ambitious of all: Two human handlers took turns walking three specially trained rat-sniffing dogs over more than a thousand miles of the island, searching for any sign of rats. They found none, and South Georgia was officially declared rat-free in 2018.

The South Georgia pipit bounced back almost immediately. Multiple successful nests were reported in the first breeding season after eradication in every treated area. Just a single year after the final bait was dropped, pipits returned to all the areas once occupied by rats. Visitors to the island after the eradication report that the sounds of singing pipits now drown out the croaks and roars of elephant seals. The species is no longer considered threatened.

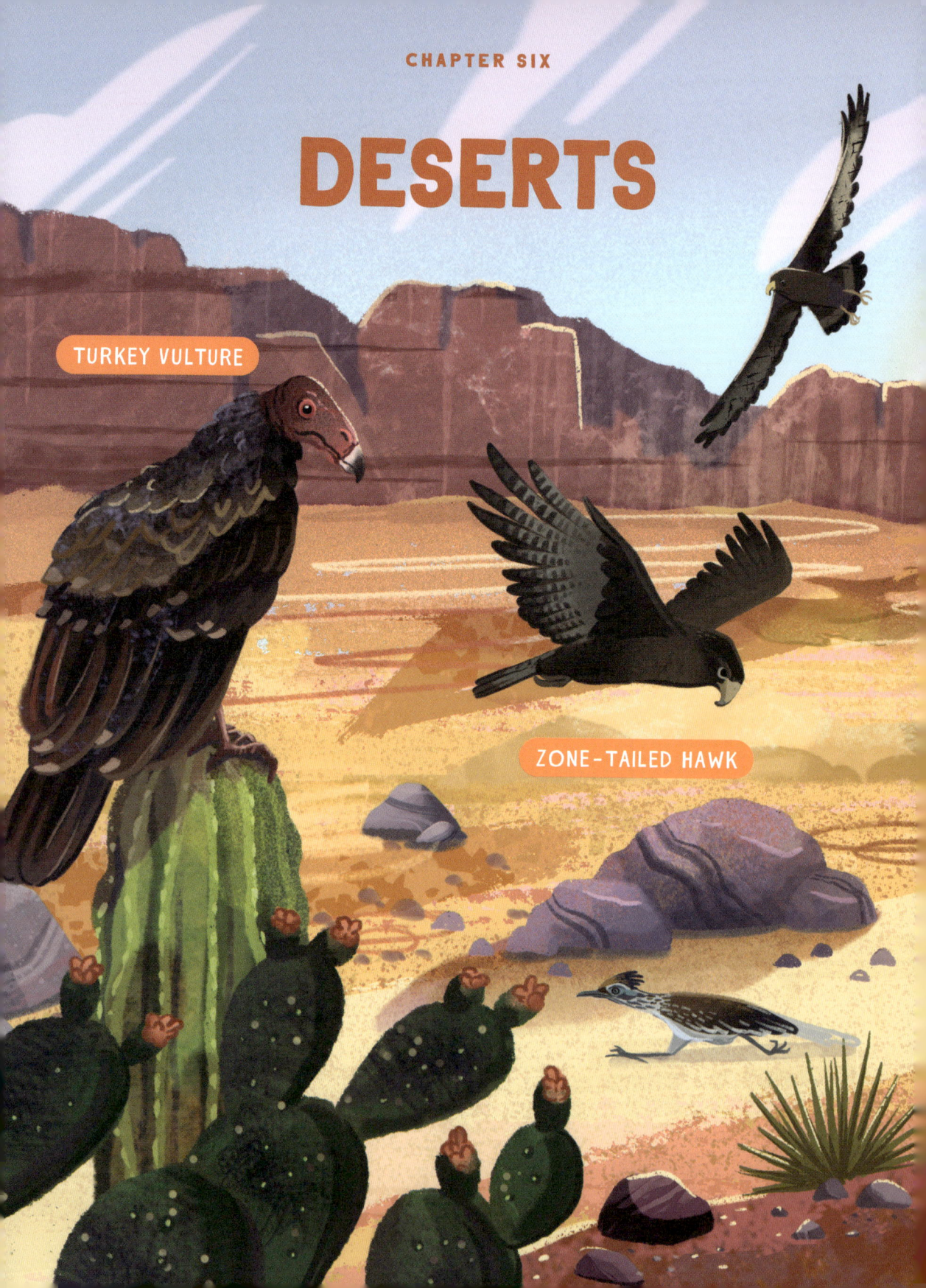

WORLD WITHOUT BIRDS

There may be no more difficult environment for any living species than the desert. **The lack of water and the resulting lack of vegetation, coupled with punishing heat and exposure to the elements, make life difficult for animals hardy enough to reside there. Yet plenty of birds do.**

Birds have to stay cool and hydrated in order to survive in the desert, and many of them have evolved special adaptations or strategies to help them out. The first might be the most fundamental of all bird traits: flight. The power of flight helps birds cover large areas quickly, permitting them to roam the desert to find what they need. Vultures soar to find meat, and cockatiels, sandgrouse, and many other birds flock to find watery oases.

But other species have developed their own tricks, from simply riding out the heat of the day in a shady spot to specially evolved organs that hold on to valuable moisture. Desert birds have to do it all to survive in the harshest of environments.

While the classic image of a desert might include big sandy dunes or cacti, deserts are actually quite varied in landscape and temperature. The term *desert* includes any area that receives very little precipitation. Ecologists agree that there are at least four different types of deserts:

- **SUBTROPICAL DESERT:** Hot and dry all year round, like the Sahara in Africa or the Sonoran in the United States.

- **COLD-WINTER DESERT:** Hot days but cold nights, like the sagebrush plains of Utah, and parts of Spain, Australia, and elsewhere.
- **COASTAL DESERT:** Found on the western edges of continents where cold ocean currents pull moisture out of the air before it reaches land, like the Atacama in Chile or the Namib in southern Africa.
- **POLAR DESERT:** The largest deserts on Earth. Though they receive relatively high levels of precipitation, the extreme weather and low temperatures create significant barriers for plants and animals, like the Arctic and Antarctic.

The harshness of deserts means there is often less of a human footprint in these places than in more hospitable environments, but with each day bringing a new battle for survival against the elements, even a little bit of change can have drastic consequences.

Staying Cool

Deserts get hot. It's kinda their thing. The hottest recorded temperatures on the planet were from deserts, including an absolutely scorching 159.3°F (70.7°C) in the Lut Desert of Iran, and a thermometer-busting 130°F (54.4°C) at Death Valley in the Mojave Desert of the United States.

Any birds living in these climates must have strategies to stay cool. One of the most basic strategies is to keep themselves thin and show some skin. Feathers can work to trap heat, keeping birds warm in cold temperatures. Think of a northern cardinal all puffed up against the cold, and now think of the opposite. Desert birds flatten their feathers down, staying sleek to keep themselves from getting too hot. And unlike the snowy owl, which has dense feathers covering its legs and feet, a desert-dwelling burrowing owl has long, bare legs that it can expose to the air to keep cooler.

Another basic desert adaptation is called *evaporative cooling*. Water cools a surface as it evaporates, and animals can take advantage. When a dog pants on a hot summer day, it moves its breath over its wet mouth, evaporating the liquid and cooling the blood vessels inside. Birds do the same thing by flapping the loose skin under their bills to move air around their mouths, a process called *gular fluttering*.

Many birds flutter their gulars to stay cool, but another bird takes evaporative cooling to a

Burrowing Owl

whole other level. Desert birds could cool a much larger area if they were able to get liquid onto their legs, but where to get that liquid? The turkey vulture uses its urine, peeing on its own legs to reap the benefits of evaporation.

Some desert birds have a simpler strategy to beat the heat: Stay out of the sun. Many desert birds ride out the heat of the day under whatever shade they can. That could be underneath a rocky overhang for Australia's chestnut-quilled rock pigeon, underground inside a commandeered prairie-dog hole if you're a burrowing owl from North or South America, or deep in a shrub if you're a saxaul sparrow from Central Asia.

Staying in the shade is even more important during the breeding season. Finding a safe spot to construct a nest in an area with few trees can be a challenge, but some birds have come up with innovative solutions. The tiny elf owl of Mexico and the American Southwest enlarges holes woodpeckers have tapped into large cacti, finding itself not just a nice place to nest but a fortress defended by prickly spines. Sociable weavers of the Kalahari Desert in southern Africa work together to build gigantic nests in trees, constructing avian apartment buildings that can house more than a hundred pairs of birds. The centers of the giant nests retain heat, and the birds can move inside during cold nights, but stay in the shade of the outer rooms during the heat of the day.

In fact, some birds actually seek out the relative isolation and safety of deserts to nest. Scientists searched for years for the nest of the Markham's storm-petrel, a small bird that spends its entire life outside of the nesting season on the open ocean. Colonies were first found in the late 1980s in the middle of the Atacama Desert in South America. These seabirds find safety in raising their young in natural holes or excavated cavities in one of the most arid—and predator-free—areas on Earth.

Finding Food and Water

The other fundamental challenge in the desert is finding food and water. To help give themselves a better chance to survive, many desert birds have evolved to become better at absorbing and retaining water. Some species, for example, have evolved extra-efficient kidney organs, which excrete almost no liquid. Others can survive with little other than the water found naturally in their foods, like berries or insects. Some desert birds get all the water they need just from seeds.

Other species need more than what they can get from seeds, but food and water are hard to find in the desert. Certain desert birds have adapted by becoming nomadic—searching far and wide for food and water and taking advantage when opportunities present themselves. Turkey vultures, for example, can soar for miles—barely even flapping their wings—searching the deserts of North America for dead animals to eat.

Watering holes are similarly unpredictable. A lack of reliable water can be particularly difficult during nesting season, when thirsty chicks are still too young to find food on their own. Sandgrouse in Africa and Asia have solved this problem by evolving special feathers on their bodies that act like sponges. When a parent sandgrouse finds a watering hole, it lies down in the liquid. The feathers on its belly absorb water and hold it until the bird can fly all the way back to its nest to give chicks a drink.

Other birds responded to the fickle nature of food in deserts by flying long distances in search of sustenance. Flocks of colorful cockatiels in Australia and pinyon jays in North America cover large areas in groups in search of nourishment—acacia seeds for the cockatiels and pinyon pine cones for the jays. The opportunism even applies to birds' mating cycles. Cockatiel mating seasons are triggered by spring rainfall, which gives some assurance to parents that their young will have food to eat.

But while food and water are challenging to find in deserts, some birds specifically travel to deserts to feed. Just as the Markham's storm-petrel makes a surprising

> In addition to hosting nesting hawks, woodpeckers, finches, and owls, the mighty saguaro cactus of the Sonoran Desert produces fruit that feeds dozens of hungry desert species.

trip from the ocean to the desert to breed, several South American flamingo species journey to the Atacama Desert in winter to feed in ephemeral salt lakes. These seasonal ponds contain tiny shrimp, algae, and crustaceans for the flamingoes to eat, but the water is too salty for many other birds—or flamingo predators—to tolerate.

Andean Flamingo

IN A WORLD WITHOUT BIRDS: CARRION CLEANERS

At first, it might not be obvious that the birds are gone from the desert. Desert animals, including birds, can be hard to spot, especially when they hide in the shade during the hottest parts of the day. Some species are also naturally scarce, making the landscape seem empty.

But eventually, you'd realize something was missing. Your first clue might be the smell.

Vultures patrol the skies over many desert and semidesert landscapes, using their powerful senses to detect dead animals to eat. The majority of vulture species around the world hunt by sight. These species have evolved large, broad wings, helping them to soar for miles with barely a movement. Their balanced flight allows them to keep a steady eye on the ground to look for meat—or, just as commonly, to follow other vultures who've already spotted a meal.

In North and South America some vultures use their sense of smell in addition to their eyes. Turkey vultures have been recorded sniffing out carcasses buried by scientists, an adaptation that helps them find food hidden under trees or grasses.

Vultures are some of the world's most important scavengers.

Their ability to cover so much ground means they're often the first scavengers on the scene, and they gorge themselves. When they find a dead animal, they can pick it clean; many vultures have evolved to lose the feathers on their heads and necks to allow them to dig deep into carcasses without getting blood or flesh stuck to their feathers. Little is left on the carcasses.

Though vultures are perhaps the most famous scavenging birds, there are many more in all kinds of habitats. Gulls flock to clean the carcasses of fish, seals, and whales that wash up on beaches around the world. Corvids—the family of birds that includes ravens, crows, jays, and others—are famous for their ability to eat almost anything, including carrion, and are found on all continents except Antarctica (where scavenging duties are taken over by skuas, snowy sheathbills, and snow petrels). Bald eagles scavenge carrion, as does the entire subfamily of raptors called *caracaras*. Dead meat is still meat, and it's nutritious food for any birds that can digest it, so long as they get to it before it rots.

Vultures and other scavenging birds are critical because dead animals become dangerous if they are left too long. Bacteria and other pathogens thrive on decaying flesh and could spread into the local ecosystem if left to rot. Vultures have special acids in their stomachs that allow them to metabolize all kinds of pathogens, including rabies, anthrax, and botulism. Other scavengers can't neutralize pathogens and may spread them to humans or wildlife.

Flies and other insects lay their eggs on dead animals, and their growing larvae feed on the flesh. Thousands of insects can grow on a single carcass, and the landscape could become inundated with flies if carrion were left to sit. Scavenging birds get to carcasses quickly and eat the meat before it becomes harmful. Without scavengers, potential health impacts for wildlife, livestock, and humans—and the smell!—would be devastating.

Birds are not the only scavengers, however, and so if birds disappeared, the populations of other meat-eating mammals would likely increase. Hundreds of species around the world would benefit, including foxes, jackals, hyenas, coyotes, polar bears, wolves, raccoons, and feral dogs. However, the fact that scavenging mammals are unable to neutralize dangerous things in the meat means that diseases would spread.

It's a lesson that some parts of the world are already experiencing. Vulture numbers are in steep decline across Africa, Asia, and Europe, in large part due to the birds being inadvertently poisoned after feeding on dead livestock that had been treated with certain medications.

The problem is particularly bad in the country of India, where vulture numbers have declined by up to 90 percent since the 1990s. The lack of vultures has caused real problems. Drinking water has become contaminated by carrion left to rot on the landscape. Populations of other animals—crows, rats, and as many as 7 million feral dogs—increased because of all the uneaten carrion now available. Feral dogs and rats bite humans and spread disease, including rabies. In fact, India has the world's highest annual death count from rabies, between 18,000 and 20,000 people each year, mostly from feral dog attacks. If vultures had been around to clean carcasses before the dogs could get there, those deaths may have been avoided.

Indian Vulture

The loss of scavenging birds would create similar problems around the world. Cleaning carcasses is a vital ecosystem service, and these birds play a crucial role in keeping the landscape clean and healthy.

HUMAN ENCROACHMENT AND THE EXTINCTION OF THE ARABIAN OSTRICH

There's nothing common about the common ostrich. It is, for starters, the largest bird in the world. From the tips of their massive toes to the tops of their small, downy heads, ostriches can stand up to nine feet tall and weigh more than 300 pounds. They may just be the most recognizable birds in the world.

The common ostrich is, for many, closely associated with Africa, but one subspecies ranged off the continent. The Arabian ostrich once lived in the deserts of the Middle East, including areas that are now the countries of Iran, Israel, Saudi Arabia, and others on the Arabian Peninsula. However, nearby civilizations flourished in the region, and the Arabian ostrich was in the path of an expanding population of humans.

Arabian Ostrich

The Arabian ostrich appeared frequently in art and culture of early civilizations in the region. It was surely hunted by early Pleistocene humans. Images of ostriches estimated at 3,000 to 5,000 years old have been found carved into rocks in modern-day Saudi Arabia. Ostriches were captured and exhibited as exotic pets in Mesopotamia during the Bronze Age around the year 1800 BCE. They were given as gifts as far away as China, their eggs were turned into fancy goblets, and they were chosen to fight gladiators in ancient Rome.

Ostriches across their range were threatened by the feather trade throughout the mid-nineteenth century and into the twentieth (see also the short-tailed albatross in Chapter 8). The dramatic puffs of feathers found on an ostrich, which could be dyed any color, were the height of fashion for women's headwear during the Victorian era. Hunters on horseback chased the birds through Arabia and North Africa to feed the insatiable demand in Europe. The demand only grew as populations shrank and feathers became harder to find.

Some believe that the only thing that saved the wild ostrich was the domestication of the bird in South Africa beginning in 1865. The birds could be raised on a farm and feathers could be harvested twice per year without killing the bird. Ostrich farms were an incredible success, and ostrich feathers were South Africa's fourth largest export in the 1880s, after gold, diamonds, and wool.

However, by the time domestication took the pressure off wild birds, it was perhaps already too late for the Arabian ostrich. Decimated by years of exploitation, the bird faced even bigger challenges. Hunters were now armed with long-distance rifles that could take down a bird from a great distance, turning the open conditions in the desert—once an aid to allow birds to spot danger—into a liability for birds with nowhere to hide. Motor vehicles, with their tireless pursuit, replaced horses and dogs and made it harder for ostriches to outpace hunters.

The Arabian ostrich was already quite rare by the turn of the twentieth century, and only scattered sightings were reported over the next few decades. A single bird was seen east of the Jordanian-Iraqi border in 1928. Pipeline workers reported shooting and eating a bird in east Saudi Arabia in 1941. Other, later, reports are unverified, and none have been reported anywhere since 1966.

The Arabian ostrich is considered the largest bird to go extinct in modern times. It joins a group of mammals—including the lion and the cheetah—to have its Arabian populations decimated by a surging human population. While the overall population of the common ostrich remains healthy throughout its large range in continental Africa, the desert specialist subspecies is gone. They evolved to survive in some of the planet's harshest places, but they couldn't survive humans.

INTRODUCED CATS AND THE SURVIVAL OF THE NIGHT PARROT

The night parrot has been called the world's most mysterious bird. It's a parrot—technically in the same family as the bright, colorful, loud birds living throughout the tropics—but different from tropical parrots in almost every way. Instead of squawking noisily through the trees looking for fruit and nuts, the night parrot lives in the remotest deserts of central Australia. Instead of bold colors, it sports a sedate green-and-yellow plumage. Instead of brashness, it hides on the ground in deep grass all day long, only emerging at night to find seeds and water.

The Western world was introduced to the night parrot in 1845 when a specimen was collected on an expedition through the vast, underexplored center of Australia led by Captain Charles Sturt. A few other scientists sought the species in the late nineteenth century. The most successful was naturalist Frederick Andrews, working on behalf of the South Australian Museum, who is believed to have collected at least twenty-two individual birds across Australia in the 1870s.

Then . . . nothing. The sightings stopped coming. The bird was presumed extinct, and no one knew why or what to do about it.

There was speculation, of course. Cats were a leading suspect. Domestic cats are, in fact, one of the leading causes of bird death in the world. In the United States alone it's estimated that cats kill between 1.3 and 4 *billion* individual birds per year. Scientists estimate that at least 175 reptiles, birds, and mammals are threatened by or were driven to extinction by feral cats on at least 120 islands.

The problem is especially acute in Australia, where feral cats are considered the single greatest threat to native wildlife. Cats have contributed to extinctions of more than twenty Australian mammals, including the rusty numbat, the desert bandicoot, the broad-faced potoroo, and the crescent nailtail wallaby. Feral cats in Australia also threaten dozens of species of birds, reptiles, and frogs. There are millions of feral cats in Australia, living in the coldest mountain peaks to the hottest deserts, and they kill millions of native animals every day. Night parrots were almost certainly on the menu.

Unconfirmed night parrot sightings trickled in for years from the remote outback, but it wasn't until 1990—more than one hundred years since the last specimen was collected—that scientists once again held the body of a night parrot in their hands, the victim of roadkill in southwestern Queensland. The rediscovery thrilled Australians and renewed interest in the species.

Even more discoveries were made after this new information came to light. In 2017, Indigenous rangers nearly all the way on the other side of Australia used a camera trap to capture images of night parrots. A better understanding of habitat needs has led to the discovery of at least fourteen different night parrot populations in western Australia, most of them on Indigenous lands and cared for by Indigenous scientists. Still, the night parrot is believed to be critically endangered.

Australia is taking action to protect native wildlife from feral cats. Different strategies are employed in different parts of the country, including cat registration, exclusion areas, trapping, and removal. It's a controversial practice, but scientists agree that invasive cat populations must be reduced if populations of native Australian wildlife, including the night parrot, are to survive.

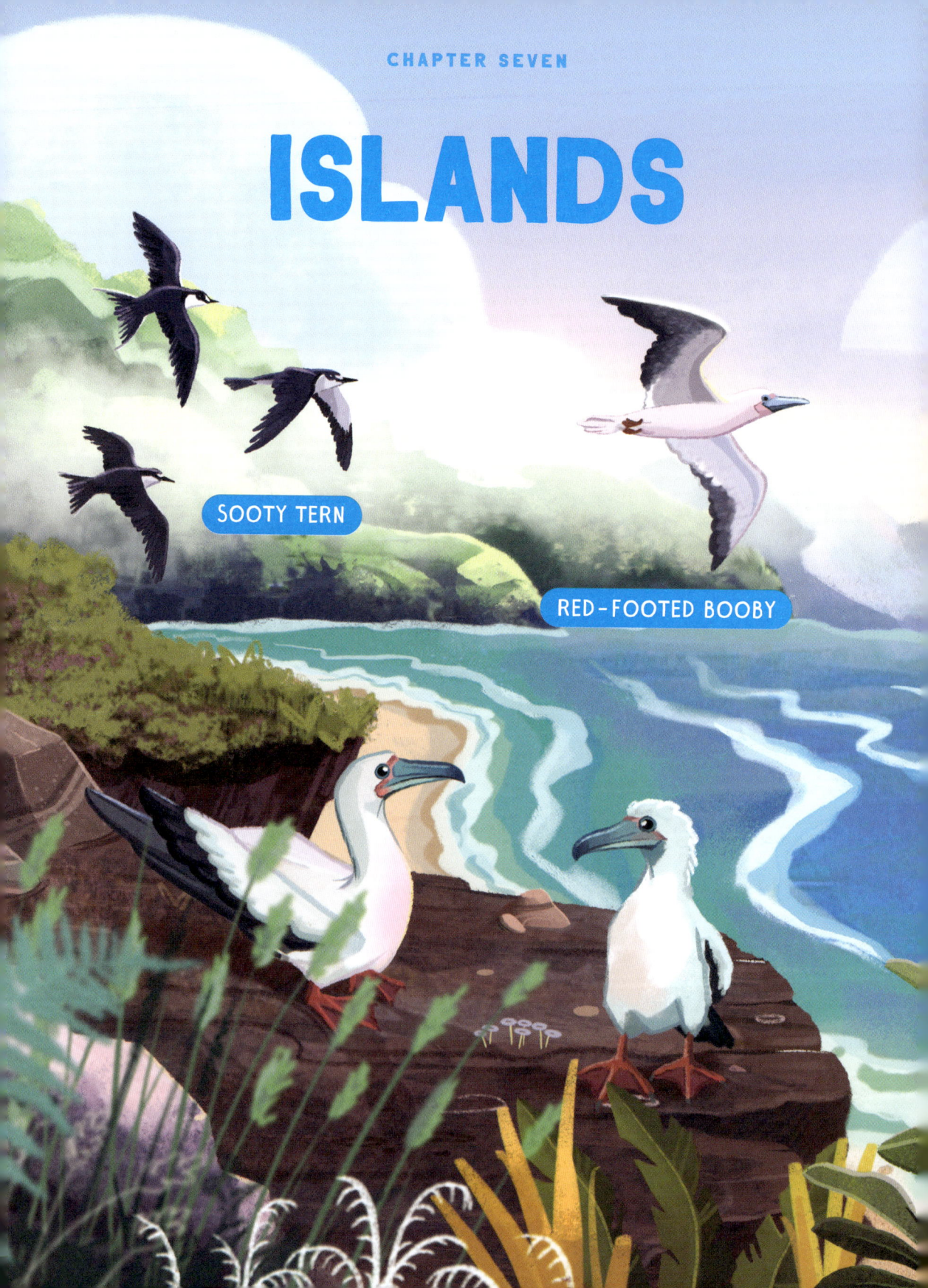

Islands are laboratories for evolution. A bird species that finds itself on an island may one day evolve to look completely different than its relatives on the mainland. Remote islands provide a different set of pressures, habitats, and opportunities for birds who end up there and drive change.

The awareness of the unusual role of islands as boosters of evolution helped us understand the concept of evolution in the first place. When the young naturalist Charles Darwin landed on the remote Galapagos Islands off the coast of Ecuador in 1835, he noticed that the local birds were similar to those on mainland South America—but different. He began to develop theories about how birds on the Galapagos arrived from South America and then changed into new species to survive in the new island environments. His theories on evolution completely changed human understanding of biodiversity, ecology, and biology.

So, how does island evolution work? Water is a major barrier for animals that can't fly or swim, like many land mammals. If volcanoes created a new island off the coast of Africa, how would a lion get there? Lions can swim, but probably not well enough to swim to an island miles offshore. It's possible that the island would simply not have lions or even any other land-based mammals. No mice or rats or squirrels or voles or cheetahs.

But animals that fly or swim *could* reach this new island. Turtles and alligators could swim there. Birds, bats, and insects could fly over. And any of these creatures that made their way to the island would find themselves in a place without mammals. A bird on the mainland that would have to fly away to avoid being eaten by a baboon wouldn't have to worry about baboons anymore. A bird on the mainland that had to compete with insect-eating mammals would have all the bugs to itself on the island. Without having to worry about being eaten by mammals or competing with them for food, birds could live differently.

Darwin understood that because birds didn't face the same pressures on remote islands as they did on the mainland, island species could grow to become different from their mainland relatives. They could evolve to better suit themselves for the particular conditions of their new homes.

It's not just the Galapagos: Many small, remote islands have their own endemic species, found nowhere else in the world.

However, these island species are especially vulnerable.

Evolving to lose mainland defenses like flight or even wariness means that species have few ways to defend themselves when predators do arrive. Plus, population sizes of island species are by definition smaller, as there is no additional territory to spread into.

Island species have been hit hard by the onset of human travel. Sailing ships looking for food found easy targets in docile island species, and the rats, cats, dogs, and other alien species released from ships disrupted fragile island habitats. In total, a staggering 80 percent of past extinctions and a third of threatened terrestrial species are found on islands.

Growing Big

What unique island birds share is not their habitat but their remoteness. Some islands are tropical, like the Hawaiian Islands or New Caledonia. Some are polar, like the South Georgia group near Antarctica. But in each case these islands were too far for terrestrial mammals to reach, presenting a very different environment for the birds and other wildlife that did show up.

Evolutionary pressures on islands are often much different from those of the mainland. One way that island wildlife often respond is by evolving to become differently sized than their mainland counterparts.

The smaller size and limited resources of islands often work to make large animals smaller. The prehistoric record is littered

ISLANDS

with dwarf versions of otherwise-large mammals that became confined on islands by rising sea levels. There were species of dwarf hippopotamus and elephant in Cyprus, for example.

Other times, the lack of mainland predators on islands allows typically small species to become larger, and so it often is with birds. The most incredible examples come from New Zealand. Apparently, no mammals were on board when the islands drifted away from Australia 80 million years ago, so birds had the run of the place. When there were mammalian predators around, many birds stayed safe by staying small so they could hide or fly. Without mammalian predators, the birds could get much bigger. The common species of swamphen in Australia weighs about two pounds, but the species that evolved in New Zealand, the takahē, is three times as large. The kākāpō, a six-pound flightless parrot only found in New Zealand, is the largest parrot in the world.

Even larger birds existed in New Zealand's prehistory. Some birds evolved to become large grazers—ecological roles that would have elsewhere been filled by large mammals like deer or buffalo.

Kākāpō

WORLD WITHOUT BIRDS

On New Zealand these were the moas, a group of shaggy, long-necked birds somewhat resembling ostriches. One moa species, the North Island giant moa, is believed to be the tallest bird that ever lived, at over twelve feet.

But where there are large grazers, there are also large predators. Moas were hunted by a massive predatory eagle, the Haast's eagle, which, at an estimated thirty pounds, was more than twice as heavy as the bald eagle. It was a mighty killer, with four-inch talons and the strength to take down moas that were twelve times heavier. The first people in New Zealand, the Māori, encountered both moas and Haast's eagles, and tell stories of the eagles killing humans.

Islands other than New Zealand produced their own massive bird species, like the dodo and red rail of Mauritius, the giant swan from Sicily, several massive goose species on Hawaii, the giant hawk

Moa

At more than three feet tall and twenty pounds, the flightless Cuban giant owl is believed to have been the largest owl that ever lived and died out 11,000 years ago.

and titan-hawk from Cuba, large owls in the Caribbean and in the Mediterranean, and others. Each of these species ruled their particular island homes, but few survived the arrival of humans.

Losing Flight

Another effect of living on an island free of predators is that flying becomes less important. On the mainland, flight is a massive advantage. It allows birds to escape from predators, find safe places for their nests, and cover long distances in search of food. On islands, though, those things are less important. There aren't predators to threaten birds or their nests, and the small size of remote islands means there aren't long distances to cover for food. Flight is hard—it requires lots of powerful muscles and lots of energy. You may as well walk.

Many bird species on remote islands have evolved to lose the ability to fly. Flightless attributes exist elsewhere in the bird world—for large, fast birds like the ostrich and certain swimming birds like penguins—but birds from many different families have become flightless on islands.

Rails are a family of thin wading birds that like to hang out in marsh grasses, and most of the world's rail species can fly. But there are more than a dozen species of rails living on remote islands that have lost the ability to fly, like the Guam rail, Lord Howe woodhen, and Inaccessible Island rail. Most of the world's ducks can fly, but not the Auckland Island teal, the Falkland steamer duck, and the Campbell teal from New Zealand. The Galapagos Islands boast a flightless cormorant. New Zealand has its famous kiwis. All of these species behave somewhat similarly to their mainland cousins (well, except the kiwis, whose closest relatives are believed to be the extinct elephant birds from Madagascar), just without needing to fly.

There used to be many more flightless island species. In fact, by far the majority of known flightless species are extinct. There were ducks and geese, flightless pigeons like the dodo and the Saint Helena dove, flightless cranes, and many more rail species. There were flightless ibis and owls and hoopoes and buntings and wrens. But they didn't survive.

> The expression "as thin as a rail" is derived from the rail family of birds, which are skinny enough to walk between dense marsh grass.

The trouble with losing the ability to fly is that you may need it if predators show up. Humans began exploring the seas in boats as early as 70,000 years ago, quickly reaching remote parts of the globe that had until then never been inhabited before. The sailors needed food, and flightless birds were easy prey. Sailors also often brought other wildlife, including predators like dogs and cats, for which flightless birds had no defense. Many island species went extinct, and many more continued to face threats in their remote habitats.

IN A WORLD WITHOUT BIRDS: EMPTY ECOSYSTEMS

If birds disappeared, many remote islands may become desolate. Birds are often the most common forest animals on islands that terrestrial mammals were unable to reach on their own, and so birds play an outsize role in how island ecosystems work. If they disappeared, their ecosystems would suffer.

Nowhere are birds more important than in New Zealand. As islands without native mammals (except bats), birds took on roles that mammals had elsewhere. There were no moles or voles to eat worms and other insects, so kiwis evolved to eat them. There were no large grazing mammals like gazelles or bison, so the massive moas evolved to take advantage of the plentiful grasses. Birds like the Haast's eagle took over the role of apex predator usually occupied by tigers or bears or other mammalian predators. Birds do everything in New Zealand: pollinate plants, transport seeds, eat insects, and more.

These roles aren't just good for the birds, they're good for the ecosystem. Grazing animals, whether a horse or a moa, help grasslands by cycling nutrients, increasing soil health, and improving plant diversity. Predators—whether apex carnivores or small insectivores—help keep populations of bugs and herbivores in check and prevent them from overrunning the landscape.

Every animal plays some role in how an ecosystem functions, and on remote islands birds play a more important role than they do elsewhere. That means that if island birds disappeared, entire ecosystems may fail.

In Hawaii, some plants have grown to become completely dependent on birds for pollination and seed dispersal. There are nearly eighty species of *Cyanea* plants in the Hawaiian Islands, and the majority of them rely on birds for reproduction. Native Hawaiian honeycreepers feed on the nectar of *Cyanea* and related plants, and some of them, like the stunning 'i'iwi, have evolved special bills to help them dip into the long flowers of certain species. The flower leaves a bit of pollen on the bird's forehead while it drinks, which is flown to the next flower to complete pollination. Honeycreepers and other native species also eat *Cyanea* fruit and spread the seeds in their poop.

If Hawaii's native birds disappeared, the plants that rely on them would also disappear. We know, because Hawaii's native birds *are* disappearing: At least seventeen of the forty-one species of Hawaiian honeycreeper have gone extinct since 1500, in addition to other native species. Hawaii's plants are suffering right alongside, with an estimated 250 native Hawaiian species so endangered that fewer than fifty of each of those plants remain in the wild.

In the absence of birds, their roles would go unfulfilled. Plants that rely on birds for pollination would die out and be replaced by those that pollinate by wind or other means.

In some places, the loss of birds is obvious. Nearly all native birds on Guam died after the brown tree snake was introduced in the 1940s. New trees stopped growing without birds to transport their seeds, and bugs formerly eaten by birds proliferated. One study found that the density of spiders in Guam was up to forty times greater than that of the nearby island of Saipan, which does not have the tree snake.

There are places where the ecosystems haven't collapsed when native birds have disappeared from islands. Instead of leaving ecological roles vacant, native birds were replaced by other animals.

Humans are responsible for all modern extinctions of island bird species, either directly through hunting and habitat loss or resulting from the introductions of mice, rats, cats, dogs, insects, plants, and other wildlife. Bursting onto islands that had existed for millennia without mammals, humans devastated the unprepared landscapes and took them over, replacing native species with nonnative ones. The outsize roles that birds played on remote islands were replaced by the mammals that filled those roles on the mainland. Grazing birds were replaced by livestock. Pollinating birds were replaced by bees and other insects. Native plants were replaced by domesticated fruit trees, crops, and nonnative flowers that didn't evolve needing to be pollinated by birds. Humans even brought in birds from other places that could outcompete native species for seeds and nesting areas. Island ecosystems were completely changed, but humans, being responsible, hardly noticed.

The disappearance of native birds has undoubtedly altered these islands, though. In parts of Hawaii, it is possible to stand in a seemingly pristine tropical forest that, in reality, doesn't contain a single native plant. Dozens of species of introduced birds and mammals transport the seeds of dozens more species of introduced plants, creating stable, functioning ecosystems in places they didn't exist until humans brought them together. The reality is that island ecosystems aren't failing, but much of what made them so unique—the birds and plants and animals that evolved to live there and nowhere else—are being lost.

ISLANDS

THE DISCOVERY OF EXTINCTION AND THE DODO

The dodo is perhaps the most famous extinct bird of all.

Dodos began their evolutionary history as pigeons. Pigeons and doves are particularly strong fliers, and many remote tropical islands have their own endemic populations. The dodo's nearest relatives are found on New Guinea, Samoa, and islands off the coast of east India. Once the dodo's ancestors landed on the island of Mauritius in the Indian Ocean, they began to adapt to their gentle new surroundings. Gone were mammalian predators that required pigeons elsewhere to take refuge in trees. Fruit and nuts were plentiful, and could easily be found simply scattered on the ground under trees. They didn't need to fly so they lost the ability to, and they grew large because the islands were safe for them to grow.

Dodo

And grow they did. The dodo stood more than three feet tall and weighed up to forty pounds—about twice as heavy as a wild turkey.

They were grayish, with lighter-colored wings, and perhaps some curled feathers at the tail. They had a strong, hooked bill and a featherless face. Their legs were strong and scaly, and they were fast runners. For millions of years the dodo lived in relative peace on Mauritius, in the middle of the ocean.

The arrival of humans changed everything. Dutch sailors arrived in 1598 and began cutting the island's valuable ebony trees. They also began killing dodos.

The birds were defenseless against the rapid and varied threats of humans. Dodos were hunted for food—they were one of the few sources of fresh meat on an island without large mammals. The dodo's forest habitat was destroyed as humans cleared land for timber and agriculture. The biggest threat to dodos may have been introduced species: cats attacked chicks, rats ate eggs, and pigs and monkeys devoured the scattered fruit and nuts that fed the dodos. The last accepted account of a dodo was from a shipwrecked sailor on a small island off Mauritius in 1662, and scientists believe the species was extinct sometime soon after.

In some ways the extinction of the dodo is just the beginning of the story. The legend of the dodo carried on, and eventually became a symbol for the very idea of extinction itself. For many people, it was perhaps the first time they wrestled with the concept. Dodos became known outside of Mauritius from paintings and other literature, most famously in Lewis Carroll's *Alice's Adventures in Wonderland* in 1865.

ISLANDS

The prevailing notion for centuries was that the dodo stood idly by as it was eradicated. Though it was included as a kind and thoughtful character in Carroll's book, it was most often portrayed in a ridiculous light, depicted as a glutton for its large size, and stupid for its fearlessness of humans. The English use of the word *dodo* to mean "stupid person" dates back to the late 1800s, and phrases like "dead as a dodo" and "gone the way of the dodo" are still in common usage.

Recently, however, the perspective on the dodo has changed, and humanity has woken up to what it lost. A forty-pound pigeon! How incredible! As one of the most famous extinct species, a reminder of how carelessly and permanently we can cause destruction. The bird is a source of pride in Mauritius—it appears on its coat of arms and on its currency—and a source of identity. Above all, the sad story of the dodo has helped spur an understanding among other island nations of the fragile existence of their own unique wildlife. On many islands around the world, that understanding came too late.

ISLAND DEVELOPMENT AND THE SURVIVAL OF THE KAGU

Islands have limited space, and so human development can quickly throw natural systems out of balance. Island birds have nowhere else to go. But humans love islands, and we're expanding our presence on islands around the world, often at the expense of native landscapes. Sometimes it leads to extinction. Other times, as with the kagu of New Caledonia, the birds are protected before it's too late.

While scientists are certain that the ancestors of the dodo were pigeons, no one is completely sure how the kagu evolved. This nearly flightless gray bird that stalks the forests of New Caledonia looks sort of like some familiar birds—a heron? A rail?—but not exactly like anything. Its closest living relative is another oddity living halfway across the globe—the sunbittern of Central and South America—but the kagu is in a family all by itself.

Whatever species the kagu evolved from, its history is similar to that of the dodo. Its ancestors made it to a mammal-free island in the middle of the ocean—this time New Caledonia, off Australia—and eventually became a completely unique new species.

The kagu is strikingly odd looking, like someone trying to describe a heron from memory. The bird has a long gray-white body standing on legs that somehow look a little too short. It walks with a start-stop motion—freezing in place for a few moments to check its surroundings before darting forward a few steps and stopping again.

People are thought to have arrived in New Caledonia around 1200 BCE, and quickly established themselves on the island. The clearing of land for agriculture and subsistence hunting are believed to have had some role in the reduction of kagu numbers in those days.

Things got worse when Europeans settled the island in the 1840s. The unusual bird was highly sought after in zoos and private collections around the world, and many were trapped and taken off the island. Agricultural development, nickel mining, and logging for native sandalwood, valued for the density of the wood and the fragrance of the bark, severely reduced the amount of native forest on the island, and today less than 20 percent remains.

Kagus retreated into the shrinking forest, where they became easy prey for introduced mammals. Pigs ate eggs and rutted up the land,

making it harder for the birds to search for worms. Rats and cats ate eggs and nestlings. Dogs, mostly hunting dogs straying from nearby settlements, are thought to be the worst threat. In one two-month period half of all radio-tagged kagu in Pic Ningua Park were killed by just two individual dogs, resulting in over 75 percent of kagu families in the area being destroyed.

Yet, despite all that, the kagu is not extinct. Conservationists on New Caledonia are clearing rats and dogs from remaining kagu strongholds and have also started captive breeding programs. Progress was slow but steady. By the mid-1990s there were around 400 kagu living just in Rivière Bleue Park. Additional birds were released elsewhere on the island, and today the overall population has grown to nearly a thousand. While that is still a very low number, it is ten times what it was a few decades ago, and a major conservation success story.

Fragile island ecosystems that evolved in isolation for millions of years were wrecked within just years or decades of human arrival, and we're still working to slow our impact. We know that with effort we can restore island ecosystems and save island birds, and we owe it to magnificent creatures like the kagu to try.

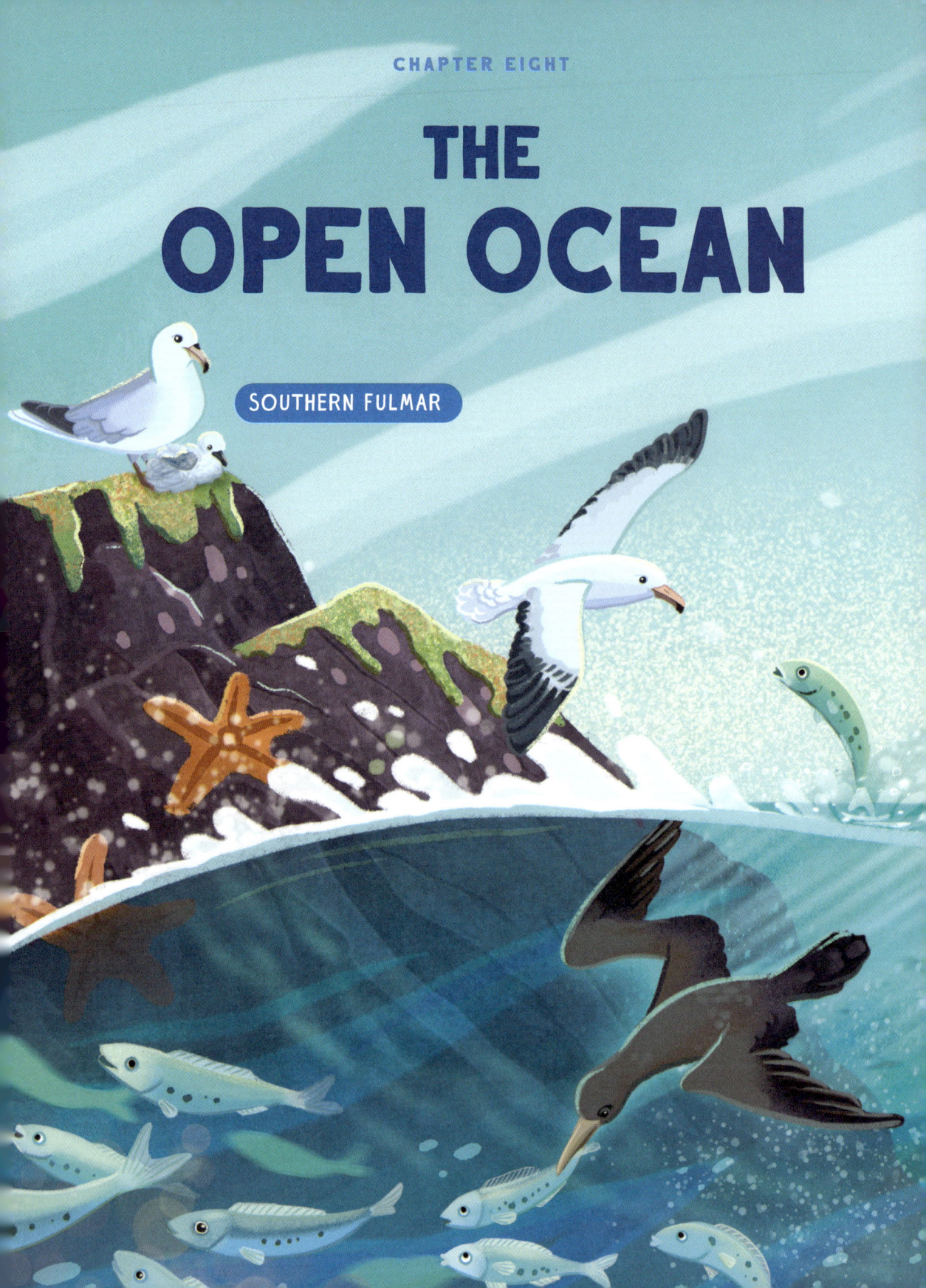

WORLD WITHOUT BIRDS

More than 70 percent of the surface of the world is covered by ocean. **It's an unforgiving and featureless habitat—there are no trees to perch in, nowhere to find shelter or shade, and food is often scarce—but birds have evolved to feel right at home.**

Seabirds are some of the most incredible and mysterious of all birds. Many have evolved special bodies to help them cover vast distances of ocean in search of food. The wandering albatross has the longest wingspan of any bird on Earth—up to twelve feet long. It's such an adept and aerodynamic flier that individuals can fly nearly 500 miles in a single day—about the distance from Atlanta, Georgia, to Washington, DC—with just a few flaps of their wings. The Wilson's storm-petrel, on the other hand, is barely the size of an American robin but flies across nearly every ocean on the planet. Though it's rarely seen from shore, there are *lots* of Wilson's storm-petrels on the ocean: they have an estimated worldwide population of 100 million.

There are many species of *pelagic* birds—ones that spend the majority of their lives at sea. These long-winged seabirds include albatross, petrels, shearwaters, fulmars, storm-petrels, skuas, jaegers, prions, terns, frigate birds, gannets, and others. All of them patrol the seas in pursuit of their favored prey—usually fish and squid. Seabirds are found in every ocean on the planet, from the tropics to the poles.

THE OPEN OCEAN

It may be tempting to hope that since seabirds are found so far from human civilization, their populations may have fared better than birds in closer proximity to humans. Unfortunately, this is not the case. Humans have severely impacted seabird populations around the world. Many species are harmed when humans arrive at the places seabirds are most vulnerable: their nesting grounds. Seabirds need to come to land to lay their eggs, and usually pick small, remote islands for their colonies. With few appropriate islands to choose from, some of these islands are crowded with thousands or millions of birds. Human disruption of these island ecosystems can have massive repercussions, as was the case for one extinct species, the Guadalupe storm-petrel.

The short-tailed albatross was also harassed at its breeding colonies. There are few creatures on Earth as graceful as an albatross in flight, but on land they are less mobile and easy prey. Albatross are also harmed at sea, and are frequently killed as accidental bycatch from longline fisheries. Yet the short-tailed albatross survived its traumatic contact with humans and is on the road to recovery. With effort and dedication, all seabirds could join it.

At Sea

How does a seabird find food on the open ocean? The sea is almost unimaginably large—thousands of miles of blue stretching from horizon to horizon, with no clear indications that food is waiting in any particular spot.

No bright red berries hanging from a branch or fat, juicy flies buzzing through the air (birds on land have it easy!)—seabird food is hidden below the waves.

But they need to eat. So how do they do it?

One strategy is to cover a lot of area, and seabirds are masters of long-distance flying. Studies have revealed that some albatross species fly an average of 600 miles per day, and certain individuals are known to fly around the entire globe in just forty-six days.

The secret, for long-winged seabirds like petrels and albatross, are bodies specially evolved to fly long distances with very little effort. Long, narrow wings are shaped to give

THE OPEN OCEAN

the birds maximum lift, allowing them to glide without flapping and conserve energy. A variety of soaring techniques help them to take advantage of small gusts of wind coming off the waves, and certain seabirds are known to cut a roller coaster of curves through the air as they travel.

Seabirds cover such distances in search of all kinds of food, including small fish, shrimp, squid, and even tiny floating animals called *zooplankton*. To locate prey on the open ocean, seabirds use powers of sight and smell.

Certain prey can be visible from the air, and some species of seabirds constantly scan the surface in pursuit of schools of fish making ripples on the surface, or, for some night-feeding species, bioluminescence— light produced in the bodies of squid and other prey. Seabirds also keep a close eye on each other, and the sight of birds dropping to the surface to feed can bring others from far away. Some species are known to fly in coordinated lines to rake across the sea looking for food, and when one finds it, they all converge. Fishing vessels are also easy to spot, and many seabirds follow ships to get floating leftovers.

Seabirds are among just a few of the world's birds that use smell to detect prey. Special tubelike structures on their bills help scan the breeze for the smells of food, including scents as faint as the

> Most albatross species mate for life, so choosing the right partner is important. Some species perform elaborate dances to impress each other, featuring dozens of moves including head flicks, bill claps, and synchronized bows.

chemicals given off when tiny zooplankton eat tiny floating plants called *phytoplankton*. These nasal tubes also help secrete salt that the birds separate from the ocean water via special glands, allowing them to drink without ever needing to find fresh water.

Once they find food, seabirds have all kinds of ways of eating it. Some, like the frigate birds, snag the food from the surface, barely even getting their beaks wet. Storm-petrels position their broad wings into the wind and use their big feet to bounce off the water's surface, plucking out bits of food as they go. Touching down with their feet is thought to help storm-petrels attract or scare prey, making them easier to catch.

Other seabirds dive right in. Fish are most often below the surface, of course, and so birds may need to go swimming if they want a meal. Some shearwaters dive underwater and use their wings to propel them to depths of up to sixty feet in search of fish. Perhaps the most dramatic divers are gannets, which spot prey from hundreds of feet in the air and then tuck their wings back and plunge like spears, beak first, into the water while barely making a splash.

Northern Gannet

THE OPEN OCEAN

Coming to Land

Seabirds have figured out how to do everything on the water except for one: lay eggs. Mating and chick-rearing require them to find dry land. For many this is the only time in their lives that they leave the ocean. Most seabirds prefer nesting locations that are on remote islands: places that are close to their feeding grounds on the water and without terrestrial predators to give them trouble.

More than 95 percent of all seabird species nest together in large colonies. Seabirds may seem scarce when they're spaced out over the massive oceans, but when they come together during the breeding season, their colonies are as crowded as cities.

Breeding birds can carpet favorite breeding islands, nesting on whatever surfaces they can manage, including sheer cliff faces. Thousands, hundreds of thousands, or even millions of birds can nest together in single colonies. The squawking and screeching of adults and young combines with the constant swirl of birds leaving and returning to the islands from hunting trips at sea to form a kind of seabird tornado. The noise, the sights—and the smell!—can be overwhelming to witness.

Nesting strategies differ depending on the location. On the barren rocks or sheer cliff faces of many seabird islands in the far north or

> Some seabird colonies host millions of birds. For example, 8 million sooty shearwaters nest on Guafo Island in Chile, while an estimated 3 million seabirds of twenty-three different species visit Midway Atoll in the middle of the Pacific each breeding season.

south, seabirds may lay eggs directly on the rock. Building a nest out of vegetation is simply not an option where there's no vegetation to be had. Sometimes cliff-nesting seabirds may surround their eggs with pebbles and guano to help keep them from falling.

On flat or sandy islands, birds may just nest right on flat ground, packed together on dirt or grass. Low-lying islands like Laysan or Midway Atoll, both part of the United States, are in this way blanketed with millions of nesting albatross, terns, and other seabirds.

Where there is both soil and elevation, seabirds may nest in burrows. Holes in the ground are perhaps the safest places for nests, as eggs and chicks are protected from predators and weather. Puffins, shearwaters, and storm-petrels all nest in burrows, typically clawed out of the soil with their feet.

The reason so many seabirds nest in colonies has long puzzled scientists, as there are drawbacks to the strategy. There's only so much space on small islands, and birds often engage in fierce competition for a particular spot or are forced to choose precarious or awkward locations. The close proximity of so many individuals promotes the transmission of diseases, and can increase the odds that chicks are injured or killed in the swarm of bodies.

Yet clearly there are benefits, or else colonial nesting wouldn't be so popular with so many species. One simple fact is that offshore islands are near food sources. Seabirds need to keep their feeding trips as short as possible if they want to raise hungry chicks, so nesting near food is critical.

Likely the biggest benefit of colonial nesting is safety. Offshore colonies are (or were, at least) often free from the kinds of mammalian predators found inland. Plus, when predators do show up, such as peregrine falcons or Steller's sea eagles, the seabirds have lots of individuals available for defense or can use their massive numbers to overwhelm and confuse predators. It's a strategy that works. There are seabird colonies in every ocean on Earth.

IN A WORLD WITHOUT BIRDS: ECONOMIC IMPACT

A world without birds would also be a much worse world for humans. We love birds, not just because they are beautiful and inspire in us a sense of joy and wonder, but also because we need them. Birds provide us with things that help us live—their bodies, their feathers, the protection of our crops and our environment. Studies have estimated that nearly all humans use half of all bird species for some purpose, whether it be for food, for their feathers, as pets, and more. Humans need birds, and we would have to make major changes to live in a world without them.

WORLD WITHOUT BIRDS

In the early 1800s the Western world discovered that seabird guano—literally the poop from birds deposited onto their nesting islands—makes for incredible fertilizer. It was an ingredient the Inca living along the Pacific coast of Peru had known for hundreds of years, but as the word of the power of guano reached agricultural centers in Europe and the United States, offshore bird poop became an international commodity. Millions of pounds of guano from Guanay cormorants, Peruvian pelicans, and Peruvian boobies was extracted from islands off Peru in the 1850s. In some places the guano was 200 feet thick. As humans do, however, we harvested the supply of guano faster than the birds could supply it, and the Age of Guano lasted just a few decades. However, guano is still harvested from islands off South America and around the world, a resource that fetches more than a billion dollars each year. Without seabirds there'd be no seabird guano, and farmers would be forced to used costly and dangerous chemical fertilizers.

Birds help farmers with more than just their waste, though. They eat insects that would otherwise eat crops. Pest control falls into a category of indirect but undeniable benefits that birds provide humans, called *ecosystem services*, which also include things like eating carrion and other waste, pollinating plants, and dispersing seeds. These services are vital for humans, but because they happen naturally, or out of our sight, birds often don't get the credit they deserve. These services are also notoriously difficult to quantify in terms of dollar value, but they would surely cost humans billions and billions of dollars to replicate.

Easily the most widespread human use of birds, though, is as food.

It's hard to know exactly for how long humans (or our prehuman ancestors) have eaten birds, but it's certainly at least tens of thousands of years. The earliest hard evidence we have are the bones of rock pigeons—yes, the same birds that today fill cities around the world—that show cut marks likely left by our cousins the Neanderthals in Gibraltar, and may be as much as 67,000 years old.

Things really picked up when humans domesticated birds, giving us a reliable source of eggs and meat without having to hunt. Scientists believe that the first domesticated birds may have been geese in China about 7,000 years ago. A few thousand years later, rice farmers in Southeast Asia began attracting wild jungle fowl, which were eventually domesticated into what is today humanity's most important bird and the most abundant bird on the planet: the chicken.

There are believed to be somewhere around 30 billion domestic chickens in the world, about four for each human. There are billions more than other domesticated animals like cows or sheep, in part because it takes many more resources and about a thousand times as much land to raise a cow than to raise a chicken. Farmers and scientists continue to find ways to reduce the cost of raising chickens, making them all the more plentiful and putting them on track to surpass pork to become the number-one source of protein in the world.

It should be noted that these innovations have not at all been good for the chickens, which, on large factory farms, are often subject to terrible conditions including overcrowding and various forms of maltreatment.

In addition to their meat, chicken eggs make up a large part of our diets, both directly and as ingredients in other foods. In all, humans are believed to consume over *1 trillion* chicken eggs per year. Chicken eggs are a vital food source, but that's not all they're used for. Eggs are also used to develop flu vaccines. Scientists inject eggs with copies of the flu virus and allow the virus to replicate over a few days, when it can then be harvested from the egg and developed into various vaccines. Millions of flu doses are given each year, meaning that chickens are saving lives as well as providing food.

Chickens aren't the only species of bird we consume. Hundreds of millions of turkeys, and millions more ducks, geese, pigeons, grouse, partridge, quail, and many others provide sustenance to humans all over the globe.

The impact of somehow losing chickens and the other birds we depend on for food is incalculable. The cost and effort required to suddenly replace what may be the most abundant protein in the human diet would undoubtedly lead to starvation and upheaval. Our entire planet would need to shift its eating habits on the fly and somehow find a replacement. It would be a nightmare. Ironically, though, the birds we depend on the most are also the least at risk of extinction. Domestic chickens are the most numerous bird on Earth, and face none of the same threats that many of the world's wild birds face. Humans have and are continuing to cause bird extinctions all over the world, but the ones we rely on the most are likely here to stay.

INTRODUCED LIVESTOCK AND THE EXTINCTION OF THE GUADALUPE STORM-PETREL

Humans eat the meat of domesticated animals like cows, sheep, and goats, but those animals need to eat, too. Across the globe, huge areas of land are cleared to create areas for livestock to graze. More than 15 million square miles of land are used for livestock production—about 25 percent of all the dry land on Earth. Millions of acres of native habitats are cleared to make way for livestock every year, and the animals themselves eat everything in their path. The impact can be especially damaging in confined areas, like islands.

The Guadalupe storm-petrel nested only on Guadalupe Island, a small volcanic refuge 130 miles off the west coast of Mexico's Baja peninsula. The birds are believed to have bred in the spring, timing their arrival to miss the breeding seasons of the island's two other storm-petrel species, Ainley's and Townsend's. Like others of their family, Guadalupe storm-petrels returned to their burrows only after dark, so as to avoid predators. It's believed that they could navigate their way to their tiny burrows in the dark via a sophisticated sense of smell.

The trouble began in the early 1800s, when seal hunters arrived on the shores of Guadalupe. Reliable sources of meat were important to sealers and whalers as they crisscrossed the globe after their prey, and so they would often stock remote islands with livestock that they could then pick up next time they were in the area. This is how reindeer from the Arctic north made it to South Georgia Island near Antarctica; pigs made it to Auckland Island, south of New Zealand, and to the Galapagos; and other mammals came to other islands around the world. On Guadalupe, the whalers left goats.

The effect of these large herbivores into ecosystems that evolved without them was, as you can imagine, devastating. On Guadalupe, the voracious goats ate almost everything in sight. Set loose onto an island with no predators, plenty of vegetation, and no competition, the goat population exploded, reaching upward of 100,000 by 1870. The goats ate everything. There was not only a disastrous decline in the number of individual plants, but an elimination of entire plant communities. Where the island was once covered in a pristine blanket of trees and shrubs—including at least thirty-five plant species found nowhere else on Earth—the goats ate it into a moonscape, covered mostly by bare and eroded soils and rocks or by weeds.

The drastic change to its ecosystem was too much for the Guadalupe storm-petrel. The last breeding bird ever recorded was seen in 1912.

Despite the devastation, there is hope for Guadalupe. The Mexican government completed the removal of more than 15,000 feral goats from the island in 2007, over the objections of the resident Mexican navy, who enjoyed having fresh meat available for barbecues. Almost immediately the island's vegetation began to recover. Scientists found that plants considered extinct or extirpated have been rediscovered—tucked into areas hidden from goats or brought over from neighboring islets—and plant species new to the island have been recorded. In areas impacted by the goats, the landscape has changed from 0 percent native vegetation to more than 50 percent in just over ten years.

There is hope that the seabirds may recover as well. While the Guadalupe storm-petrel is gone, the Ainley's and Townsend's storm-petrels have small but stable breeding populations on tiny nearby islands, Islote Negro and Islote Afuera, free of predators. If vegetation continues to return to Guadalupe, stabilizing soils and providing hiding places for burrows, and work to remove cats and mice continues, there's a good chance these seabirds may return to the main island. Guadalupe is healing, though it will forever be without some of its former native residents.

THE FEATHER TRADE AND THE UNLIKELY SURVIVAL OF THE SHORT-TAILED ALBATROSS

People have been adorning themselves with bird feathers for thousands of years. Fancy feathers were important parts of ceremonial garments across the world, from eagle feathers in Indigenous American garments to bird of paradise feathers among early inhabitants of Papua New Guinea.

THE OPEN OCEAN

In Europe, the wearing of feathers was an indication of style, culture, and civility in the 1500s, and as the popularity of feathered headwear grew, so did the demand for birds. Great crested grebes, a resident of freshwater lakes across Europe and Asia, were shot to near extinction for their chestnut-brown ear feather tufts. Ostriches were extirpated from the Middle East and Northern Africa, and were only saved as a species once feathers from farm-raised ostriches became available. The feather fad continued for hundreds of years, and eventually crossed the Atlantic. Experts estimate that, at its height at the end of the nineteenth century, around 200 million individual birds—mostly egrets, with their long head plumes—were being killed for their feathers each year.

Nesting seabirds were easy prey. While albatross are graceful and powerful in the air, they are clumsy on land. Like many birds that evolved in isolation, seabirds don't fear humans and are easily approached. Plume hunters exploited seabirds at nesting colonies around the world, killing millions of terns, gulls, kittiwakes, albatross, petrels, and other species. Some, like the short-tailed albatross, very nearly went extinct.

It is believed that the short-tailed albatross

was at one time the most abundant albatross species in the North Pacific. It may also be the most beautiful. When soaring, its black wings stretch more than seven feet across a white body, topped with a golden head and neck and a pink bill. It's colorful, as far as albatross go.

When breeding season comes, short-tailed albatross head for just a few small islands, mostly off Japan and islands south of Japan. The largest is the remote volcanic island of Tori-shima, alone in the Pacific about 373 miles south of Tokyo. The island had been uninhabited by humans, save for the occasional shipwreck survivor, until enterprising plume hunters came ashore in 1887.

Each dead albatross yielded about a quarter pound of feathers, and in some years the Tori-shima colonies produced up to 350 tons of feathers—the equivalent of 2,800,000 birds. Scientists believe that, in all, more than 5 million short-tailed albatross were killed on Tori-shima alone. By the early 1930s, the birds were gone.

But then, in 1951, the director of the island's weather station took a walk onto the backside of the island and—there they were—ten adult albatross. Where they had been for so long may never be known. Albatross spend years at sea before they reach maturity, but not twenty years. They'd simply eluded everyone, and were just clinging to life.

Conservation measures were taken seriously this time, and the population began to grow. The first eggs were laid in 1954, and in the years since, the global population has risen to an estimated 4,200 individuals, with 3,540 of those birds on Tori-shima. The short-tailed albatross had miraculously survived to a point where humans could finally control themselves, and their population could now recover.

Serious threats remain. The first is Tori-shima itself. It's still an active volcano, and an eruption during the breeding season could destroy all its birds. As of 2014, some birds had been moved to other traditional colonies, including 650 birds on two islands in the East China Sea and ten birds on the Ogasawara Islands. The birds have made attempts at nesting on other islands, including the famous colony on Midway Atoll, part of Hawaii, and there is hope for additional expansion.

Short-tailed Albatross

Freshwater ecosystems are critical to the survival of thousands of bird species. All birds—all living things, including humans—need fresh water to survive, but for many it's more than just a drink: It's a home. Lakes, ponds, rivers, streams, marshes, swamps, and other freshwater ecosystems play a central role in the lives of thousands of bird species around the world.

Ducks, geese, and other waterfowl live right on the water. They spend most of their lives floating on the surface, relatively safe from predators and with all the food they need in the water below them. Wading birds and shorebirds use a variety of techniques to hunt along in fertile shallow waters, from dipping their bills into the mud to find morsels underground to waiting patiently to stab at an unsuspecting fish. Marsh birds like rails, gallinules, moorhens, and others use tall wetland vegetation to stay hidden from predators.

Yet freshwater ecosystems are threatened around the world, and many of the species that rely on them are threatened, endangered, or already extinct.

The impacts can be broken down into four categories: complete destruction of the ecosystem; alteration of the physical habitat;

changes to the water quality or quantity; and changes to the species makeup.

Wetlands and other freshwater ecosystems can be completely destroyed by developers, who fill them in to make buildable land for other projects. They can also be destroyed by taking in so much water for humans to drink or to water crops that the areas just aren't wetlands anymore. Alteration takes many forms, such as damming rivers to make reservoirs, channeling rivers, and removing their natural shape and flow.

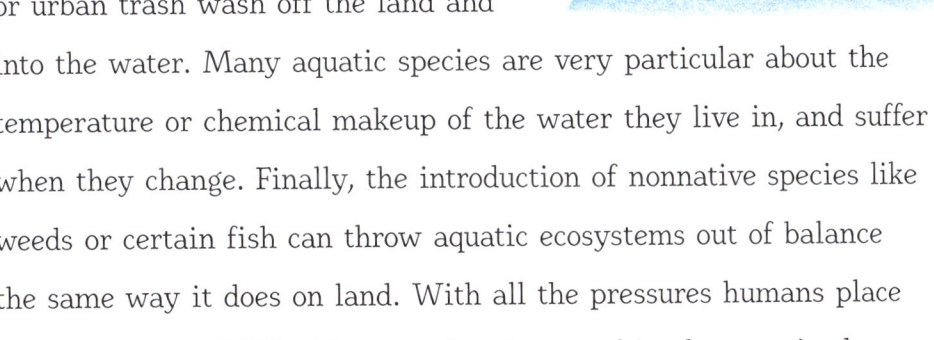

Water quality changes when pollution from industrial waste is dumped into a river, or when agricultural chemicals or urban trash wash off the land and into the water. Many aquatic species are very particular about the temperature or chemical makeup of the water they live in, and suffer when they change. Finally, the introduction of nonnative species like weeds or certain fish can throw aquatic ecosystems out of balance the same way it does on land. With all the pressures humans place on freshwater wildlife, it's a wonder that anything has survived.

On the Water

A large and familiar group of birds from the Anatidae family are typically found floating around on fresh water. Better known as waterfowl, this family includes ducks, geese, and swans.

WORLD WITHOUT BIRDS

Waterfowl have figured things out pretty well for themselves. They get most everything they need right on the water. There's food swimming around underneath them, and they can spot danger coming from a long way away and either dive underwater or fly to safety. It's a way of life that works; waterfowl are found almost everywhere on Earth, from the tropics to the poles.

They've evolved some special features to help them succeed in a watery world. Of course, the ability to swim is crucial. Waterfowl don't have much need to use their toes for clutching prey like a raptor or perching on branches like a songbird, so their toes have evolved into paddles. The toes are connected by skin, forming a webbing that helps push the birds through the water. The foot spreads out when the bird pushes its foot backward, then folds together when it comes forward to reduce drag.

Webbed feet are such an asset to swimming birds that they've evolved independently in several bird groups, not just waterfowl but also loons, gulls, cormorants, puffins, and others. Even other kinds of animals that live in the water, like frogs and beavers, have evolved webbed feet.

FRESH WATER

Agility in the water is critical for waterfowl to feed. Waterfowl food, whether it's aquatic invertebrates, submerged vegetation, or other morsels, is all underwater, and the birds need to reach it. For some, just reaching down and grabbing it is enough. Some ducks, like the northern shoveler, simply filter water across their bills and pluck out the edible bits. Many ducks and swans feed by "dabbling": tipping their backsides into the air and sticking their heads into the water, reaching out with long necks to pluck vegetation off the bottom of the lake.

Others dive down for their food. Most waterfowl simply dunk themselves underwater and use their large feet to propel themselves to the bottom to collect plants or mollusks. Other freshwater birds are after more elusive prey fish. Ducks called mergansers have evolved sleek, powerful bodies with long necks to snag fish, and serrated bills to hold on to slippery prey. Nothing underwater is safe from waterfowl.

> Though most species of waterfowl nest at the water's edge, some look for a more secluded setting. Wood ducks, mergansers, and goldeneye ducks all nest in tree cavities excavated by woodpeckers.

One quirk of life on fresh water is that it may not work all year long. Waterfowl living in areas with cold winters can't feed on frozen lakes, and so millions of waterfowl migrate to find open water. It can be one of the natural world's most amazing spectacles: long strings of honking geese moving overhead, or lakes suddenly covered in ducks just flown in from the north. Migration is also a physical achievement, as some waterfowl are among the fastest-

and highest-flying birds. Bar-headed geese, for example, have been recorded flying at more than 23,000 feet during their semiannual trips over the Himalayas.

In the Mud

Not all freshwater birds live in the deep end, however; life is also good in the shallows. Hundreds of species worldwide have evolved to wade along the edges of lakes or in other wetland areas, including herons, egrets, ibis, spoonbills, bitterns, and more.

Reddish Egret

Herons, egrets, and bitterns are predators. They use stealth and speed—and sharp bills like javelins—to spear fish and other aquatic prey. Different species have different tactics to catch their food. Some are patient. Great blue herons and pinnated bitterns are among the many species of wading bird that stand and wait for a fish to swim close. When one does, the bird launches itself bill first into the water to spear the prey. Others, like the reddish egret and little egret, are more active, dancing back and forth through the water. It's thought that their movements help stir up prey, making the fish easier for the bird to see.

Some herons get even sneakier. A few herons, including the green heron of North America and the worldwide striated heron, use their bills to place bits of food like insects or twigs on the surface of the water. Fish are attracted to the floating food, and come close enough for the heron to pounce.

> The famous shoebill of North African marshes uses its massive mandibles to snag larger fish than most any other heron, some more than thirty inches long.

Another heron, the black heron of Africa, tricks fish using its famous "umbrella" technique. The bird stands in the shallows and stretches its wings over its head, forming a sun-blocking canopy. Fish are attracted to the shade thinking that it's a safe place to hide, and the heron picks them off.

Fish aren't the only items on the menu in the shallows. Another group of birds, the shorebirds, hunts for prey burrowed in the mud. Sandpipers, turnstones, curlews, plovers, and many others stalk the water lines along both freshwater and saltwater coasts, probing the mud with their bills in search of aquatic invertebrates.

Sometimes a dozen or more shorebird species can be found along a single mudflat, so how do they coexist while sharing the same habitat? The answer is that each species has evolved to find food in its own particular way. Some species, like the long-billed curlew, use their extended beaks to reach worms and crabs more than eight inches underground. Dowitchers and yellowlegs have somewhat shorter bills, and probe for prey just a few inches beneath the mud. Plovers and sandpipers use their comparatively stubby bills to snag prey from at or just below the surface. Thanks to each species finding its own niche to exploit, thousands of individual shorebirds of many species can coexist on a single stretch of shallow water while ensuring there's plenty of food to go around.

IN A WORLD WITHOUT BIRDS: CULTURAL LOSS

Birds have always held a special place in human culture. Their omnipresence, their beauty, and the fact that they rarely pose any danger to humans—unlike some mammals, reptiles, and bugs—means that we have a fondness for birds that we don't have for many other animals. Birds feel like friends.

Our lives would be so boring without them. Can you imagine? Losing birds would leave a void in our souls, and perhaps sever one of our last strong connections to nature.

The first thing you'd notice in a world without birds would be the silence.

In general, being a bird means it's easier to be loud. A bird's ability to fly away from danger means they don't always need to be as cautious as other animals. There's a reason that mice, for example, aren't brightly colored and don't sing. Mice are food for a

FRESH WATER

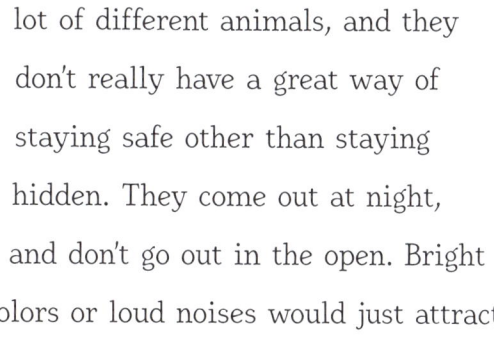

Blue Jay

lot of different animals, and they don't really have a great way of staying safe other than staying hidden. They come out at night, and don't go out in the open. Bright colors or loud noises would just attract unwanted attention.

Thanks to their flight abilities, birds can be brightly colored, and they can be loud. Mornings during breeding season can be deafening, as dozens of bird species belt out their own particular songs to proclaim their territories and invite potential mates to take a tour. Even outside of the breeding season, there are always some birds making noise in the forest: a crow cawing, a jay squealing, or a nuthatch honking. If birds disappeared, the silence would be overwhelming.

Bird sounds reaffirm our connection to the natural world.

Wherever you are—in a city or a desert or a frozen wasteland—birds are there, too, reminding us that humans aren't the only creatures on the planet. Their omnipresence helped birds become an important part of different human cultures throughout history and across the globe.

WORLD WITHOUT BIRDS

Images of owls were painted alongside ibex and aurochs in French caves 15,000 through 35,000 years ago in the Upper Paleolithic. The oldest piece of artwork known from China, dating to more than 13,000 years old, is a small carved bird. Depictions of dozens of bird species remain in ancient Egyptian artwork, as well as on frescoes from ancient Greece and in the oral traditions of Indigenous Americans. The *tinikling* dance from the Philippines was inspired by the careful walk of certain rail species. Eagles, falcons, and other birds of prey have been used to represent the strength of armies or nations since at least the Achaemenid Empire around 550 BCE, all the way through modern symbolism of the United States, Mexico, Zambia, Egypt, Montenegro, and others.

Birds were inspirations for countless poets, painters, and songwriters working in the English language. In the nineteenth century, Percy Bysshe Shelley found beauty in birdsong with "To a Skylark" and John Keats explored nature and mortality in "Ode to a Nightingale," while Edgar Allan Poe was trading on the inherent eeriness of crows in "The Raven." Alfred Hitchcock also used mobbing birds to instill fear in *The Birds*, while the Beatles drew parallels to the

American Civil Rights Movement in "Blackbird." There are countless examples.

In the English-speaking world, birds are symbols of love (doves and swans), freedom (swallows and eagles), wisdom (owls), beauty (peacocks and nightingales), and power (eagles and falcons). Someone who quickly learns something new takes to it *like a duck to water*, while someone who doesn't might be a *dodo*. A child moving away from home leaves an *empty nest*, but when they're gone they're *as free as a bird*.

The point is, birds are part of who we are. They're in our speech, in our art, and in our psyches. So what would happen to us if they disappeared? Would Americans have to remove the eagles from the backs of our quarters? Would the Toronto Blue Jays, St. Louis Cardinals, Philadelphia Eagles, and Atlanta Falcons have to change their team names?

We'd certainly lose richness and beauty in our lives. Various studies have shown that people are spending less and less time outside, and that our connection with the outside world is as tenuous as it's ever been. For many, especially those in urban areas, birds may be the last regular contact they have with wild animals. Though we haven't always treated them well, birds have always been there. Their disappearance would mean the end of a relationship that has spanned millennia, and would be a disaster from which we may never recover.

POLLUTION AND THE EXTINCTION OF THE COLOMBIAN GREBE

Human activities often leave harmful materials behind, better known as *pollution*. Pollution of the air, water, and land can linger for years, disrupting ecosystems, harming wildlife, and posing long-term threats to biodiversity. Sometimes it causes extinction.

Grebes are a family of small water birds that have evolved a number of special adaptations to help them chase fish and crustaceans. Their legs have moved toward the back of their body, acting like an outboard motor to push the bird forward. Grebe feet aren't webbed like those of other ducks and waterbirds, but feature thick, lobed toes, like their feet were squashed. Each toe can move independently, helping the grebe swim quickly and with great agility.

There are about twenty species of grebe on the planet today. There used to be another, a species found only in the wetlands near what is now Bogota, Colombia: the Colombian grebe. Little is known for sure about the particulars of the Colombian grebe's

Colombian Grebe

habits or appearance, but scientists believe it had unremarkable gray-and-white plumage during the nonbreeding season but a dashing black-and-chestnut during mating season. Male breeding plumage featured bold tufts of yellowish feathers spraying out from behind the eye.

The Colombian grebe inhabited what was once a massive wetland complex. Erosion and tectonic activity high in the mountains created natural dams across rivers flowing out of the mountains, in turn forming chains of large lakes. Over time these lakes filled with sediment, turning them into a massive patchwork of swamps, marshes, and shallow lakes. Such a wet and fertile place became a haven for wildlife, including at least twenty endemic species. High up in otherwise fairly barren surroundings, these wetlands were an oasis.

But the same things that made these high Andean wetlands popular with wildlife were also attractive to humans. The verdant and productive land has been home to humans for more than 12,000 years, eventually becoming the sprawling city of Bogotá. The landscape was changed drastically to fuel the region's growth, leading to an avalanche of change.

The Colombian grebe's habitat mostly disappeared, and what remained was severely degraded. Pesticides used on farmland washed into the water, along with industrial waste and urban runoff. The rivers and water systems that ran through the region and its cities became conduits for human waste and the chemical by-products of new industries rising up in the area. Water pollution has all kinds of harmful impacts on birds. Their food becomes contaminated. Their habitat deteriorates as pollutants kill vegetation and change water chemistry. Polluted water can physically harm birds, interfering with their feathers, skin, and eyes. The Colombian grebe had nowhere else to go.

In all, the rapid, unplanned urbanization of the Bogotá region was too much for the Colombian grebe. By the 1940s, the only reliable place to see the species was on a single large lake north of the city, Lake Tota. They were believed to be abundant on the lake in 1945, in groups of ten to thirty. Their numbers dwindled in subsequent years: 300 total birds were seen on the lake in 1968, then just a single record from 1972. An observer in February 1977 saw one or maybe two birds. It was the last time anyone would see a Colombian grebe alive.

AGRICULTURAL CONVERSION AND THE SURVIVAL OF THE CRESTED IBIS

Humans have converted half of the habitable land on Earth—more than 19 million square miles—into agriculture. For every square mile of grassland, forest, and scrubland, humans have cut another square mile to plant crops to feed ourselves and our livestock. Every bird that occupied those habitats has had their livable area reduced. Some didn't make it; others are just hanging on.

FRESH WATER

Ibis were made for the mud. They are wading birds, closely related to herons, egrets, and spoonbills. They typically forage on muddy shorelines and wetlands, using their long, thin bills to probe for crustaceans and other invertebrates hidden in the soil. One species, the crested ibis, used to range across eastern Russia, Japan, and China. It's a striking bird: bright white with a bare red face and long red bill. In flight, the bird shows soft orange underwings. And, of course, atop its head is a tousle of long feathers: its namesake crest.

The crested ibis was revered across its range, yet it was under assault. Wetland habitat was converted into rice paddies. Ibis flocked to the flooded fields to eat crustaceans, but farmers believed they were eating the rice and killed them on sight.

Crested Ibis

WORLD WITHOUT BIRDS

Farmers began using much stronger pesticides in the early twentieth century to improve their rice harvest, killing off much of the arthropod prey in the paddies. Ibis starved. Farmers also changed the way they managed paddies, drying them out in winter instead of keeping them wet or converting them to wheat or other dry crops. The ibis was running out of habitat.

The species disappeared from much of its large range. They weren't seen in China after 1964. The last wild sighting in Japan was in 1974; in Russia it was 1981. Many believed the species to be extinct.

But, luckily, it wasn't. An ornithologist named Liu Yinzeng led a party into a small region in the heart of China called Yangxian County in 1981, searching through rice paddies for any sign of crested ibis. In late May the party struck gold: two breeding pairs of birds and three chicks. These seven birds represented the entirety of the world's wild crested ibis population. But they also represented its hope.

The miraculous discovery in Yangxian brought a conservation movement into life. Logging was prohibited near where the birds were nesting, and hunting for the birds was banned, along with the use of pesticides in their favored paddies. Nests were patrolled and guarded during the breeding season, and nearby rice paddies were kept wet in the winter and even stocked with fish. A crested ibis nature reserve was established in 1990 in nearby Shaanxi to hold a growing population. Some fledglings were taken from the wild and bred at the Beijing Zoo, a practice that has been successfully replicated in other zoos near Yangxian. By 2010 there were five captive populations in China, with a total of about 600 birds.

The effort has expanded to other countries. Numerous reintroduction programs to Japan's Sado Island, using chicks from captive birds and even an early transplant from China's successful program, failed. But Japanese conservationists didn't give up, and another donation of birds from China eventually resulted in the first wild crested ibis nest in Japan in more than forty years in 2009. The fragile population continues to grow.

CHAPTER TEN

WHAT CAN WE DO?

WHAT CAN WE DO?

There are tons of things you can do to protect birds. This is a book about hope, remember? For every sad story of extinction presented here, there's also a story of people saving a bird—recognizing the harm we were causing and stopping it before it was too late. All those people who are saving birds are just regular people who've stood up and taken action, whether it's Rachel Carson, who rang alarm bells about pesticides and helped save the bald eagle; Neiva Guedes, who founded the project to save hyacinth macaws; Liu Yinzeng, who rediscovered the crested ibis; or countless others. Each of these people started on their own, but their work grew into movements.

Maybe you'll be able to start your own environmental movement. Maybe you could join one already underway. Maybe you could start smaller. All these efforts work, and there are ways for everyone of any ability to help protect birds.

Let's start with some big ones.

You could get a job protecting birds.

Lots of people have careers protecting birds. Here are a few options:

• **You could become an ornithologist.** Ornithologists are scientists who study birds. They work all over the world, in every habitat. Some work in the field, studying birds in the wild, taking samples and measurements, and gathering data. Other ornithologists work to analyze and communicate that data, as teachers or professors.

- **You could become an environmental lawyer.** Laws like the Endangered Species Act provide critically important defense for birds and other wildlife. Lawyers help to get environmental laws written and passed, and they represent the interests of wildlife against those accused of breaking environmental laws.

- **You could become a rehabilitator or veterinarian.** When an individual bird becomes sick or is found injured, it is up to veterinarians or the staff of a rehabilitation center to nurse the bird back to health and release it back into the wild.

- **You could become an environmental advocate.** Many environmental organizations have staff who work to promote certain policies that protect birds. Environmental advocates raise awareness of the plight of birds and ways to help them, and work with lawyers and elected officials to pass laws that protect birds.

- **You could become a professional bird guide.** Bird guides work for tour companies and take paying birders on trips to show them birds. Some guides travel all over the world; others stick to a particular state or area. Being a bird guide means sharing your love and knowledge of birds with others.

- **You could work at an organization that protects birds.** There are lots of groups out there dedicated to the protection of birds,

WHAT CAN WE DO?

and they have jobs for people with all kinds of skills. They need people to raise money, design posters, maintain the buildings and the grounds, sell merchandise, organize volunteers and campaigns, and much more.

You could join a group.

You could also volunteer and make a real difference helping birds without making it your full-time job. There are groups all across the country and the world that are looking for passionate people to help out. That could mean volunteering to count birds for a nearby chapter of the National Audubon Society, assisting in a beach cleanup with a local environmental group, or simply joining a group that you feel is most aligned with what you believe in.

There are countless environmental groups that are worth joining or donating to. Some are local or state-based; others are national and international. Look around and figure out which groups are the best fit for you. Here are some well-respected groups for you to consider:

• **National Audubon Society.** Founded in 1905, the National Audubon Society is one of the oldest wildlife conservation groups in America. There are more than 500 local Audubon chapters across the country, and they work to educate people and protect birds. *www.audubon.org.*

- **American Birding Association.** The American Birding Association is the only nationwide organization in America that caters to recreational birders, and they work to promote birdwatching and protect birds and their habitats. *www.aba.org.*

- **Cornell Lab of Ornithology.** Part of Cornell University in New York, the Cornell Lab of Ornithology works to interpret and conserve the Earth's biological diversity through research, education, and citizen science focused on birds. *www.birds.cornell.edu.*

- **American Bird Conservancy.** This nonprofit organization works to support policies that protect birds and their habitats, with a particular focus on pesticides, habitat loss, building collisions, and invasive species. *www.abcbirds.org.*

One thing that environmental groups often ask you to do is write to your representatives about a particular issue. This can sometimes feel pointless, like there's no way that one person's letter can make a difference for a powerful politician. But trust me.

Write to elected officials and tell them that you want them to protect birds.

WHAT CAN WE DO?

It's the job of elected officials to listen to the people that elected them. They may not do what those people ask, but they need to know that there are people who want to see birds protected. Our officials should know when we're disappointed in them for not protecting birds, and they should know when we're happy that they did something to protect birds. You can write, you can call, you can schedule a meeting. It may seem that they won't want to meet with you, but they will! Senators, representatives, city counselors, mayors, governors—they all need to know that the people who elected them want to see birds protected, and it's up to us to tell them.

Finally, there are other actions you can take.

There are proven and effective ways to help birds without leaving your backyard or your block.

- **Plant native plants.** Birds eat lots of things, but insects are the most important food source for baby birds. Insects eat plants, and many of them eat only specific plant species. Nurseries and greenhouses will sell you plants from all over the world, but local insects may not be able to live on them, and then birds won't have those insects to eat. Planting native plants gives birds the food they need to raise their babies.

WORLD WITHOUT BIRDS

- **Give them food and shelter.** While they're no substitute for the natural food found in native plants, bird feeders are a great way to supplement the diet of your backyard birds while also giving you a chance to see them. In addition to birdseed, consider mealworms, orange halves, suet, and other food. Birdhouses and bird boxes are also helpful, giving birds a safe place to raise their young.

- **Keep cats inside.** As you've read, invasive species like cats can be a major problem for birds. In some places, like Australia, cats have escaped and established massive feral colonies in the bush. But even pet cats kill birds and other wildlife. Keeping your pet cat indoors is the best way to ensure that local birds are safe.

- **Turn your lights out, and make your windows safe.** Glass windows, believe it or not, can be a major threat to birds. Birds accidentally fly into windows when they're tricked by the glass reflecting sky or habitat. Up to 1 billion birds per year die this way in the United States alone. Light can make the problem worse by attracting and disorienting birds. There are many products and strategies you can use to make your windows safe for birds, and turning your lights out during migration helps keep birds on track.

- **Encourage your parents to drink shade-grown coffee.** Coffee is an important crop in many parts of Central and South America, where billions of birds spend all or part of the year. Some coffee plantations cut all vegetation except the coffee plants, destroying lots of bird habitat in the process. But coffee can also be grown in the shade, meaning that farmers can keep tall trees for the birds while still getting a healthy crop.

- **Use less plastic.** Humans have produced billions of tons of plastic waste, and plastic pollution can be found almost everywhere. Plastic takes hundreds of years to decompose, and before then it poses a threat to wildlife, including seabirds who mistakenly eat floating plastic thinking it's food. Recycling, using less plastic, and supporting policies that reduce the use of plastic straws, Styrofoam, plastic bags, and other single-use plastics can help.

It's not too late to save birds. Extinction is real but so is our will to fight it. The more we understand the world around us and the more we recognize the impact we're causing to the planet, the better chance we have to avoid losing more species forever. Please: Learn, listen, and act.

The future for all birds—and our entire planet—is at stake.

NICK LUND

Nick Lund is a nature writer who mostly writes silly things about birds on social media when he should be working. He is the author of *The Ultimate Biography of Earth*, and his writing on birds and nature has appeared in *Audubon* magazine, *National Parks*, Slate.com, *The Washington Post*, *The Maine Sportsman*, *The Portland Phoenix*, *Down East* magazine, and others. He is a graduate of the University of Maine School of Law and worked in federal energy policy in Washington, DC, before returning to Maine with his wife and son to work for Maine Audubon.

ASIA ORLANDO

Asia Orlando is a visual artist, author, and illustrator with a master's degree in art and design. Her portfolio spans book illustration, advertising, packaging, and games, with clients including Disney, National Geographic, and leading publishers. Asia has illustrated more than ten books and created the illustrated novel series Mugo's Universe as both author and illustrator. She also founded Our Planet Week, an environmental art movement, and shares behind-the-scenes videos on her YouTube channel.